ENGINEERING & ARCHITECTURE

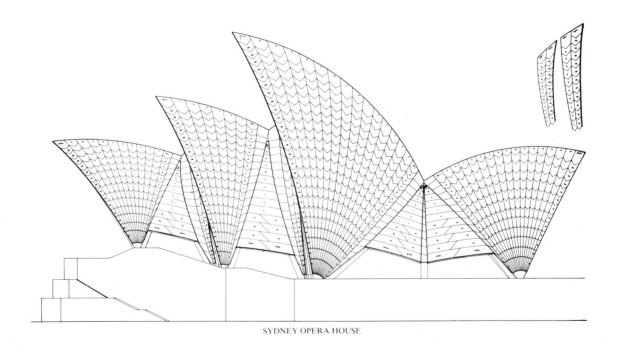

SYDNEY OPERA HOUSE

Published to coincide with an exhibition,
The Great Engineers, commemorating the 150th anniversary
of the founding of the Royal College of Art, London

TERRY FARRELL PARTNERSHIP, EMBANKMENT PLACE

ENGINEERING & ARCHITECTURE

ISAMBARD KINGDOM BRUNEL

WE WOULD LIKE TO THANK PROFESSOR DEREK Walker of the Royal College of Art for his invaluable contribution to this publication, as well as his wife Jan Walker, who provided immensely helpful research. We are also grateful to the Rector, Jocelyn Stevens, and the staff at the RCA, whose 150th anniversary is being commemorated with an exhibition on *The GreatEngineers*. The exhibition, which celebrates the important engineering achievements of the past century and a half, and raises important questions about the relationship of engineering to architecture, was the starting point for this Profile.

Many organisations and individuals have lent us valuable material: we are especially grateful to Felix Samuely & Partners for supplying contemporary photographs of the Skylon at the Festival of Britain, and to Clive Wainwright of the Victoria and Albert Museum, who lent us rare Delamotte lithographs of the Crystal Palace, hand-coloured to show the original Owen Jones colour scheme.

MICHAEL HOPKINS. SCHLUMBERGER FABRIC ROOF

Editor: Dr Andreas C Papadakis

First published in Great Britain in 1987 by *Architectural Design*
an imprint of the
ACADEMY GROUP LTD, 7 HOLLAND STREET, LONDON W8 4NA

Architectural Design Profile 70 is published as part of *Architectural Design* Volume 57 11/12-1987
Distributed in the United States of America by
St Martin's Press, 175 Fifth Avenue, New York 10010

ISBN: 0-85670-932-8 (UK)
ISBN: 0-312-01555-0 (USA)

Printed in Great Britain by E G Bond Ltd, London

INTRODUCTION

ROBERT STEPHENSON IN COMMITTEE, CONTEMPORARY PAINTING, c1850

Engineering and architecture are two sister professions separated by a morass of mutual misunderstanding. While the engineer may feel that the architect's creative perception stops at the boundary of structural logic, the architect often wishes that the engineer were better able to appreciate the aesthetic and formal implications of construction. This divide has existed as long, in fact, as there have been modern architects and engineers – since the 19th century when, in spite of

the advances of the industrial revolution, a general suspicion of material and technological development took root among the populace and grew. The very existence of two separate professions both dealing with specific aspects of the building process is evidence of society's inability or unwillingness to absorb the values of technology.

However the notion that creativity is the sole preserve of the architect is challenged by Professor Edmund Happold in his contribution to this issue. Saying that a world which sees art and engineering as divided is not seeing the whole, he suggests we re-assess the received wisdom that art necessarily expresses the individuality of man and promotes cultural evolution: '...if truth be told, ideas of history and art can entrap. It is technology that is creative because it frees the scene, by opening up new opportunities.' The engineer's craft, says Happold, is intensely creative, at its best extending people's vision of what is possible and giving them new insights. But engineering design, unlike architecture, does not follow a visual fashion or style: it is a technological idea, with its own aesthetic that is more likely to relate to a natural than to a historical precedent.

Happold gives examples of this principle in numerous projects, including his own work with architects such as Frei Otto. These buildings demonstrate a willingness to learn from nature in which, says Happold, 'structures have to be totally appropriate and mistakes become extinct': an aviary in Munich, for instance, is made using crimped steel mesh to resemble a spider's web.

He also discusses the application of a structure not only to

resist forces but also to actively amend the existing environment, as in a design for a covered city in the Arctic, where the quality of light – nature's most powerful source of energy – is improved by the use of new laminates which transmit more of the spectrum than glass.

However the best-known embodiment of the technological idea as an aesthetic in its own right probably remains the Crystal Palace. Robert Thorne's essay, 'Paxton and Prefabrication', deals not with the conventional story of how Joseph Paxton saved the day for the commissioners of the Great Exhibition in 1851, but instead looks at one of the more problematic aspects of the building's history: at how an ideal method of component construction which had been rapidly developed to a peak of efficiency faded from view, never fulfilling the potential it seemed to offer.

Thorne examines two principal issues: the fact that a modular system, however flexible, is not tailored to a single site, and can seem clumsy in many locations; and the question of whether an industrial method can produce an acceptable architectural aesthetic.

The answer to the second question would be a resounding yes if we were to judge solely by the Delamotte illustrations reproduced in this issue. These rare lithographs, produced just after the Crystal Palace was moved to Sydenham, were hand-coloured to show the original Owen Jones colour scheme, and convey the dazzling impact of the perspective down the nave.

More examples of visual art of engineering are given by Francis Pugh in 'Design, Engineering and the Art of Drawing',

L TO R: OVE ARUP & PTNRS. ENGINEERS/DEREK WALKER ASSOCIATES. ARCHITECTS. 'WONDERWORLD', CORBY, *c*1987-91, THEME BUILDING SITE AND MODEL

in which he traces the origins and development of the language of drawing from its roots in plans made by master masons in Graeco-Roman times through Vitruvius to today, with illustrations drawn from the fields of architecture, engineering, shipbuilding, surveying, and map- and instrument-making.

Much later than Paxton, another engineer who produced some of the outstanding buildings of his time – the functionalist era – was Owen Williams, who wrote a great deal on the territorial wranglings between engineers and architects:

I do not believe that an architect as an architect can collaborate with an engineer as an engineer ... You have the opposition of two philosophic ideas ... you can either maintain practicality, carry it to the extremist point. With a philosophical basis you will in this way produce the finest form of art, that is to say art is the capacity to do a job, having regard to every condition. Practicality is a method of achieving the effect without making the effect a method of achieving itself. On the other hand you have the doctrine that by effect, conscious effect, you can deliberately achieve beauty. To my mind this is very similar to a man who sets up in life and says 'I shall be a very beautiful character' and you say to him 'Be honest first and if you are honest you will be beautiful, but do not attempt to be beautiful and dishonest.' And if you think of architecture and engineering, one trying to be practical and the other trying to say 'We have a God-given mission to be effective', these two things are actually opposing doctrines which cannot collaborate.[1]

The divide between the two professions is a recurrent theme in the interview between architectural critic Charles Jencks and engineer Jack Zunz of Ove Arup. In addition, Jencks and Zunz discuss Arup's involvement in many of the seminal architectural projects of the past three decades, such as the Sydney Opera House, Pompidou Centre, and Hongkong Bank.

Arup's are also the engineers on two Terry Farrell projects in London – Lee House and Embankment Place – that are presented along with a number of other recent projects in which the engineering of the building plays not a secondary role but rather is expressed as an integral part of its aesthetics: Norman Foster's Médiathèque in Nîmes, Frei Otto's Diplomatic Club in Riyadh, and ABK's Cummins Engine Plant.

The engineering of architecture is also the subject of the essay

L TO R: THE IRON BRIDGE, COALBROOKDALE, *c*1779; RICHARD TURNER, ENGINEER/DECIMUS BURTON, ARCHITECT, PALM HOUSE AT KEW, *c*1848

L TO R: DEREK WALKER ASSOCIATES, ENERGY PAVILION, 'WONDERWORLD', CORBY, 1986, ENERGY PLAYGROUND MODEL AND CROSS SECTION

by James Gowan, who says that 'if a scrupulous distinction were made between architecture and engineering, it would be that one is concerned primarily with art and the other, utility. When one activity invades the territory of the other, it does so at considerable risk.' Gowan illustrates this risk by pointing out inconsistencies between the rhetoric and practice of contemporary architects as diverse as Richard Rogers and Quinlan Terry.

There is clearly, then, a need for greater understanding between the professions – a recognition that, in many cases, architects and engineers design more efficiently *together* – an argument put forcefully by Derek Walker in his Introduction to *The Great Engineers* book and exhibition celebrating the Royal College of Art's 150th anniversary, an event with which this issue of *AD* also coincides.

Throughout its history, the Royal College of Art has not always acknowledged the reason for its formation as a design establishment, which was to forge and maintain links with industry and to encourage an element of design within engineering by absorbing it into the process as a whole, rather than, like surface stylists, just applying a little polish to an object of overwhelming mediocrity. The climate of today's market does not warrant the mutual exclusivity of design and manufacturing at any level or scale and, because of industry's great need for invention, design, packaging and marketing, it is important we celebrate designers who really understand the process – intelligent building, corporate identity, appropriate use of materials and high levels of environmental comfort. The environment should not be a public lottery overseen by the blind. It demands concentration on quality and consistency.

Notes

1 Quoted from 'The Architect as Engineer: Sir Owen Williams' by Frank Newby and David Cottam in *The Great Engineers*, edited by Derek Walker, Academy Editions, London, 1987

L TO R: FREI OTTO, GARDEN CENTRE, MANNHEIM; ARNI FULLERTON, 58° NORTH PROJECT, CANADA *c*1983

SIR JOHN SOANE. 13 LINCOLN'S INN FIELDS. LONDON. 'THE DOME IN 1810' (DRAWN BY GEORGE BAILEY)

JAMES GOWAN
The Engineering of Architecture

I K BRUNEL, CLIFTON SUSPENSION BRIDGE, c1864

'But the magnitude of that achievement would soon have been eclipsed and forgotten did not every detail remind us of it, by reflecting Brunel's infallible eye for proportion and his sense of grandeur. His exquisite sketches of the architectural detail of tunnel mouth, bridge, or viaduct, of pediment or balustrade remain to reveal, not Brunel the engineer but Brunel the artist at work.' *LTC Rolt* **on Isambard Kingdom Brunel.**[1] **If a scrupulous distinction were made between architecture**

and engineering, it would be that one is concerned primarily with art and the other with utility. When one activity invades the territory of the other, it does so at considerable risk. In the catalogue of the 1987 Soane exhibition at the Dulwich Gallery, Michael Graves tells us that:

My long vault of the Sunar showroom from side to side is nothing like Soane's. My vaults become more Roman than his, by virtue of going from side wall to side wall. He didn't do that. He made an Attic building by these wonderful tiled vaults. They are so strange – strange in a wonderful sense. You hear all kinds of stories about whether they're structural or not, or phoney. In a way it doesn't matter but one of the features of the Dulwich vaults is that one hasn't seen them before, they are inventive and they don't seem quite right. Soane makes us look again at that Roman vaulting system which he employs at other places as well.[2]

This commentary is illustrated with an interior photograph which features flat-arched beams and clerestoreys topped with a low-slung lantern, and much glare. The effect could be said to be the contrary of what Soane stage-managed. Soane directs one's attention downwards by an orchestration of shadows and bright planes. In Graves' interior the lantern holds the spectator's gaze and the notion that this construction is more Roman than Soane's is only fleetingly entertained, for Graves' arrangements display the semantics of an attachment to a rigid frame.

Much of the architecture of Sir John Soane has about it a terrible simplicity with, in contradistinction, overlays of considerable refinement. The great recurring notion derives from

Rome, the pantheon and the hypogeum: both prototypes of the most splendid and awesome kind. The paradox is that Soane made them work for most day-to-day activities and inevitably, the chief component of his architecture was the arch. This obsession resulted in a display of great serial invention, not unlike Mies' excursions with the Chicago frame. Syntactically, the two architects could be said to be both simple and complex. In the event, both proved difficult to follow.

Three years after the end of the 1793-1815 war, Soane was asked by the Surveyor-General to advise on the design of the new Parliament-funded churches. His response, a model of clarity and common sense, ran thus:

That the interior of the churches, to be within the compass of an ordinary voice, should not exceed in length 90ft and in breadth 70ft, that the square and parallelogram are the most economical forms. That the structure as respects the walls should be of brick, and no greater quantity of stone used than is required to assist their construction, or to render the exterior characteristic, and for the requisite pavements.

A hint of incompatibility between architecture and engineering is latent in the comment:

That the gallery in small churches be sustained by iron pillars, but in those of large size their supports should be partly of stone and continued to the roof, and should it be objected that the use of iron alone has not sufficient character and appearance of stability, it may be enclosed in the manner best adapted to prevent obstruction.[3]

But sometimes the architecture of Soane belies the straightforwardness of the text. A model of the interior of an unbuilt church[4] places an arcosolian vault overhead, instead of underground, and the clerestorey carries a corbelled arcade of a type that might be expected around a courtyard, shock tactics that one has come to associate with Le Corbusier; the house perched on *pilotis*, the garden on the roof. In the Council Chamber for the Freemasons[5] each wide flat arch is carried, not on a solid wall or abutment, but on two decorative pilasters flanking a sash window with a fireplace underneath. This conjunction is ambiguous; amusing and disquieting, adroit and perverse.

When he described the Lloyd's Building at a recent and lively RCA talk, Richard Rogers was at pains to correct misinterpretation. 'The outcome was not a matter of novelty. Its assembly was sustained by a good deal of sense and more than that, a link with the past and a precedent .... the mediaeval cathedral.' He was referring to the structural exposure of this style of architecture, and the proposition he was making appeared to be that the new and the old were kindred and true in their nakedness. It is a point of view which the eye finds difficult to accept and is only a convenient historical analogy, if one is prepared to

the fundamentalists Krier, Terry, Adam *et al* and observed that the key words were 'authority', 'discipline' and 'tradition'. Apparently, structure was referred to by Quinlan Terry when he advised against cavity walls in favour of the stout and solid. It is good advice, up to a point, as solid walls have a nice simplicity about their make-up. Asking as little as possible from the British building industry is a sensible strategy. Walter Segal, part architect, part engineer, had the measure of that with his light, ephemeral structures perched on the surface, not within the good earth. What the fundamentalists do not face up to is that, operationally, classicism was creaking very badly at the start of this century. Stone walls laced with steel and cramps ran counter to the ground rules of masonry construction, and massive walls encroached upon the floor space so esteemed by developers.

But there have been periods when architecture and engineering have been disposed in sweeter accord. The stadium at Wembley, the Regent's Park pool for penguins and Boots Nottingham Factory are as stylistically adroit as any of the buildings of the new white architecture. The mention of style blurs any discussion on engineering, but it exists not only to confuse. Brunel could not have bettered the graceful lean

FELIX SAMUELY & PNTRS. ENGINEERS/POWELL & MOYA. ARCHITECTS. 'THE SKYLON'. FESTIVAL OF BRITAIN. *c*1951

jump over 20 volumes of Ruskin. To the sage, architecture was structure enhanced by decoration and hand-craft, hardly an engagement that Lloyd's can claim to be concerned with, even peripherally. Indeed, Rogers is on record as saying that his early experience with on-site improvisation, a newspaper damp-course, turned his attention and allegiance to component construction.

Lloyd's and Beaubourg have their roots in the 1950s at the Architectural Association – Archigram notions of fun, and fairgrounds and their stylish structures ... good looks, economy, and speedy prefabrication. One did not make the link then that these pleasure-buildings were not peasant art but the work of versatile engineers who kept a low profile. Not so long ago, British TV took Cedric Price and Archigram Inc on a 'Tutti-Frutti' trip to Paris where they had a light-hearted time pinpointing the bits of Beaubourg that they had put on paper in times past. I remember having an extended argument with Charles Jencks about morality in art and he held the opinion that it did not exist. On authorship, it was simply a matter of who got there first and he was probably right.

Mark Swenarton wrote recently of a symposium[6] staged by

mechanics of the Clifton Bridge, yet had trouble with the pylons whose bulk had to be given a form and, surprisingly, he chose to make them Egyptian in appearance; thus making them look lighter and more graceful. Art is often concerned with deceit and one presumes that mathematics is not. Anyway, that was how artists of the 1930s and those at the Bauhaus, in particular, saw engineering – as rational, enviable and objective. Speaking of this period Gropius says that:

the object of the Bauhaus was not to propagate a style, system, dogma, formula or vogue, but simply to exert a revitalising influence upon design. We did not base our teaching on any preconceived ideas of form, but sought the vital spark of life behind life's ever changing forms.[7]

Even so, when Leslie Martin arranged his book[8] on the furnishings of the English flat, with Sadie Speight, it was principally about style and a monolithic, brave-new one at that.

The engineer of this period, or perhaps I should say the architects' engineer, was Felix Samuely. Small, not impressive in manner, a little irascible and none too dexterous with the English language, he had taught a generation of young architects at the Architectural Association very well indeed. In the year of 1950, the talents of Philip Powell and Hidalgo Moya were joined to win the competition for the Skylon, a cigar-like structure held up by wires and steel pylons, not Egyptian in style, indeed severely purposeful. The design started off as a horizontal slim balloon, filled with helium and held to the earth by two wires running parallel, one from each tip. Samuely was impressive in action, particular and pernickety; if a beam needed 50mm bearing, then that was it. Konrad Wachsmann distinguishes between refinement in engineering and pragmatism with a comment he made in his book on the Crystal Palace:

> However, Paxton had achieved a better adaptation of the statical loads, for the thickening and tapering of the members he used correspond more closely with the forces acting than do Eiffel's parallel lattice girders, with their two-dimensional components.[9]

Technical experimentation is more of a necessity than a fad. A miracle was needed to rearrange the shambles of World War II, and science and the new men described by C P Snow seemed to offer a dynamic salvation. Apparently, World War I had similar consequences. Maurice Casteels comments upon these in his splendidly illustrated survey of *The New Style* printed in 1931.

it seemed. This mechanical preoccupation and its general application to a variety of uses continued into the school building programme which started, with great promise and elegance, in rural Hertfordshire at Cheshunt. These initiatives extended into Powell and Moya's housing at Pimlico, then rising from the ground in bright yellow brickwork. Heated by surplus energy from Battersea Power Station, the enterprise and its engineering lay unseen in a tunnel below the river and, if less dramatic than that of the Brunels at Blackwall, it was very characteristic of the spirit of the time.

Notes

1 L T C Rolt, *Isambard Kingdom Brunel*, Longman, p 141

2 Giles Waterfield, *Soane and After*, Lavenham Press, 1987, p 92-3

3 Arthur T Bolton, *The Works of Sir John Soane*, St Lukes Printing Works, p 91

4 *ibid*, p 90

5 *ibid*, p 116

  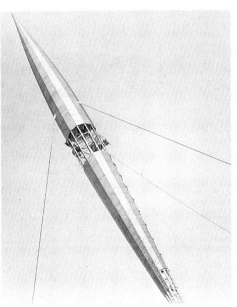

FELIX SAMUELY & PNTRS. ENGINEERS/POWELL & MOYA. ARCHITECTS. 'THE SKYLON'. FESTIVAL OF BRITAIN. c1951

The War it was that brought into being these problems that made a change essential. The War altered the whole situation, psychologically, as we all know and from the practical point of view as well ... a more sober architecture arose, an architecture of straight lines whose keynote was utility, an architecture that excluded imitations of past styles.[10]

In 1950, by book and broadcast, J B Priestley had given warmth and optimism to the cold calculations of socialism and, on the island site in front of the Victoria and Albert, the Science Museum was sponsoring an experimental building to demonstrate the effects of the Coriolis force. The engineer was Colonel Hazelhurst who specialised in fairground structures and the trials took place in a large roundabout, screened off for the purpose.

The Festival itself was dominated and enlivened by engineering gestures; the Skylon, Hungerford Bridge and the Dome of Discovery – Ralph Tubbs' masterwork with Freeman Fox. Leonard Manasseh won the open competiton for the restaurant with the novel idea of using an off-the-peg agricultural Dutch barn; prefabrication and romanticism – the best of both worlds,

6 *Building Design*, Morgan Grampian Press, March, 1987

7 Walter Gropius, *The New Architecture and the Bauhaus*, Faber and Faber, p 24

8 J L Martin and S Speight, *The Flat Book*, William Heinemann

9 Konrad Wachsmann, *The Turning Point of Building*, Reinhold p 24

10 Maurice Casteels, *The New Style*, B T Batsford, pp 20-21

James Gowan is a practising architect and Senior Tutor in Architecture at the Royal College of Art. He was a member of the design team at Powell & Moya Architects that worked on the Skylon for the Festival of Britain in 1951.

FREI OTTO, ARCHITECT/BURO HAPPOLD, ENGINEERS, MUNICH AVIARY, WEST GERMANY, *c*1980, DETAIL

EDMUND HAPPOLD
A Personal Perception of Engineering

I K BRUNEL, ATLANTIC STEAMSHIP

Professor Edmund Happold, a leading engineer designer, articulates the aesthetics of engineering design, tracing the creative spirit that is an essential component of the practice and theory of engineering from the work of the pioneer industrialists of the 18th century to what he sees as the technological idea that informs the rapidly changing innovative nature of contemporary practice which embraces the imaginative use of lightweight structures and new flexible materials.

Everyone knows the harshness of nature; how mistakes become extinct. Yet the unique characteristic of man is that his reason and imagination have enabled him to develop and adapt to his environment. And history shows how technological development has liberated him. At first it was the discovery of a type of wheat which gave an abundant yield, achieved by ploughing, sowing and reaping, which led to the possibility of settlement and the building of permanent structures. Today there is the exploration of the deep – the conquest of space. Yet in response there has been continuous concern at such advances. Fear of change. Fear of consequences. Technology, especially since the time of the industrial revolution, has been seen as a dehumanising force to be resisted.

Art has been seen as the civilising counterbalance to these advances. Art is seen as expressing the individuality of man and promoting cultural evolution. It is appropriate that this number of *Architectural Design* has its roots in the Royal College of Art's 150th anniversary since the College has its origin in a report on Arts and Manufactures which, in 1836, recommended the formation of a School of Design in which the 'direct application of the the arts to manufactures should be deemed an essential element'.

But technology (defined by Galbraith as 'the systematic application of scientific or other organised knowledge to practical tasks')[1] – or engineering if you prefer that word – obviously cannot in itself be a bad thing; it is all a matter of how we use it. And a world which sees art and engineering as divided is not seeing the world as a whole. These days often only those people with an arts training are said to be creative. But, if the truth be told, it is technology that is truly creative because it opens up genuinely new opportunities. Historic ideas of art and culture can entrap. It is technology that frees the scene.

Throughout history there has been a succession of turning points, achievements by engineers which represent a new conception of nature. These turning points reveal why engineering can be so intensely satisfying, for it is, at its best, an art grounded in social responsiblity.

Industrial archaeology, ie the history of engineering, is now culturally acceptable and certainly influences modern fashion. Yet as an engineer designer I represent an approach to design whose roots are not dependent on visual precedents. I am referring to engineering design as a technological idea, with its own aesthetic.

The engineering profession as we know it today developed to serve the non-conformist industrialists of the 18th century, who were midwives to the industrial revolution. These men, predominantly Quakers, were barred by dint of their nonconformity from the established universities and professions. They found creative opportunities within their limited possibilities by turning to inventive industry. And because they believed in the equality of mankind guided by individual conscience, they backed humanistic management. They had very broad longterm aims. They coped with persecution by forming close bonds with their fellow industrialists and these interrelationships and the pooling of ideas and information facilitated the development of the industrial revolution.[2]

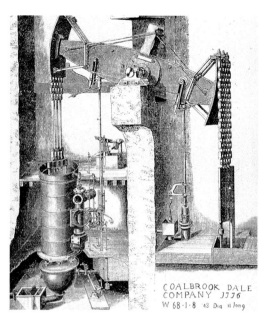

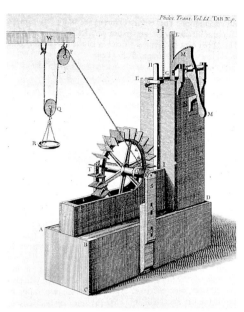

L TO R: NEWCOMEN, FIRST STEAM ENGINE; JOHN SMEATON; JOHN SMEATON, AN EXPERIMENT

The founders of our profession were creative mechanics. In 1712 Newcomen manufactured his first steam engine. In 1759 Smeaton carried out a classic study on water power. In the 40 years that he worked as a consulting engineer, Smeaton regularly used the word 'professional' to describe himself. Each morning he was employed on a time basis to consider problems and design schemes. But the fact that he saw his scientific studies as the basis of his work is described by his daughter, Mary, who wrote 'his afternoons were regularly occupied by practical experiments, or some other branch of mechanics'.[3] This interest in the scientific study of the sources of power in nature, together with the performance of materials, represents one aspect of a technologist's body of knowledge. The other aspect is in the development of construction methods; the organisation of work. Some of the achievements of the other accredited founder of the civil engineering profession in Britain, James Brindley, well illustrate this. In the construction of the Bridgewater Canal (completed 1769) he popularised the use of 'puddled clay': a mixture of readily available sand and clay which his workmen would tramp into the bottom of the canal with their boots to provide an impervious yet flexible lining.

Engineering technology really took off in 1760 when two foremen at a Quaker ironworks in Coalbrookdale produced cheap iron using coal, not wood, as fuel. It was the beginning of the era of making materials with creative potential from non-renewable resources, such as metals and fossil fuels.

In 1779 the world's first iron bridge at Coalbrookdale was complete. In 1801 the first steam carriage was built by Richard Trevithick. In 1826 Telford's great suspension bridge over the Menai was completed. George and Robert Stephenson built the first effective steam railway and I K Brunel's steam ships bridged the Atlantic. Technology transfer started early: every American knows the portrait of Whistler's mother – but his father came over to Britain to learn railway engineering from the Stephensons and not only returned to pioneer railways in America, but also went to Russia and started the construction of the Trans-Siberian railway. These engineers initiated the staggering development of technologies which have, in the last 130 years, so changed the world.

As a structural engineer, or technologist, I acknowledge that my way of thinking about the world falls into what Persig designates the 'classical mode'. You may remember that in his

L TO R: WORKMEN CREATING SAND AND CLAY LINING FOR BRIDGEWATER CANAL; IRONWORKS AT COALBROOKDALE

L TO R: JAMES BRINDLAY; T TELFORD. MENAI SUSPENSION BRIDGE. COMPLETED 1826

book *Zen and the Art of Motor Cycle Maintenance*[4] he describes those who see the world primarily in terms of immediate appearance as thinking in the 'romantic mode' and those who see the world primarily in terms of underlying form as thinking in the 'classical mode'.

His interesting example is of Mark Twain, who wrote in lyrical terms about the Mississippi River until he went to train as a river boat pilot. He gained a deeper understanding of the river through learning its science, but in the process it lost its original magic for him. Persig argues that the romantic mode tends only to develop existing forms, whereas the classical mode is capable of producing originality.

Most engineers would see themselves as falling into the latter category. Their craft is intensely creative; at its best it is art, in that it extends people's vision of what is possible and gives them new insights. But the aesthetic produced is 'bare', it may not have been seen before, and it is more likely to relate to a natural rather than historical, precedent. This is what I mean by engineering design as a technological idea as distinct from a visual style or fashion.

Perhaps the best known example of this is the Crystal Palace, produced not as an 'art object' but because it worked. Prefabrication and organisation of plant and labour, together with the significance of iron and glass as building materials, were demonstrated to the world. But to the art/architecture establishment of the time, notably Ruskin, the Crystal Palace was anathema. He advocated a return to the style of Gothic Mediaevalism.[5] It is also interesting that although fire-proofed cast-iron structures were produced only two years after the erection of the Iron-bridge at Coalbrookdale, it was a quarter-of-a-century before they were used in architect designed buildings.

Certainly in Britain there is a belief that technology must be tamed and controlled, largely by imposing standards of visual beauty which comply with criteria set up by those who have studied the arts.

65 years ago Roger Fry, amongst others, was complaining that the British were in love with ancient art and too little interested in the work of their contemporaries.[6] For my part, I sometimes wonder whether we were not almost fatally damaged by the refugees from Hitler's Europe. Those who were gifted designers, frustrated by the reactionary environment here, went on to the USA and revolutionised design there. The art critics and

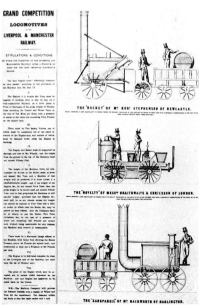
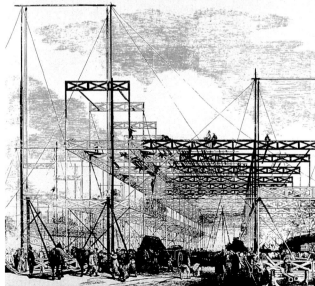

L TO R: WORLD'S FIRST IRON BRIDGE. COALBROOKDALE; GEORGE AND ROBERT STEPHENSON. FIRST STEAM RAILWAY; CRYSTAL PALACE. UNDER CONSTRUCTION

L. TO R: CONCRETE SHELLS, NERVI AND CANDELA; BUCKMINSTER FULLER, AMERICAN PAVILION, MONTREAL EXPOSITION, *c*1967

historians, discovering our conservative, class-rich environment, stayed on to become famous and powerful in our establishment.

But technology is about change. It is concerned with the development of useful objects or processes which change our lives. It does this in response to people's aspirations or is restrained by people's fears; in this it relates to the arts. What it does must obey the laws of nature, which is why it uses science to examine behaviour. Technology is the making of things while science is the explaining. So the roots of engineering are in nature.

Everything in the built environment has been achieved by technology. Every single man-made object in the world is the product of technology and traditionally the modern built environment divides into structures and machines.

The relationship between structures and machines is extremely interesting and there are many aspects they have in common. The most efficient use of materials, or perhaps energy, is a major one. Yet the energy in structures, which are seen as essentially permanent, is in the production of the materials and in the construction. Machines on the other hand, are designed to convert energy efficiently – into motion, heat, messages and the like. Structures are the steady parts of the system, machines the dynamic and the two are entirely interdependent. You cannot have an aeroplane without an airport, a generating turbine without foundations and a building, a chemical plant without a structure, cities without drains, power, etc, even a plane without engines and a frame. However, in detail structures and machines are often completely different.

But does our society really understand and recognise the great contribution engineers have made to the modern world, and are still making today? Often, I think, the answer is 'No'.

Perhaps it is not surprising that the public, with its need to 'individualise' success, fixes its approbation on the package designer whose pencil co-ordinates and markets a design. In the design of machines, such as washing machines, computers or cars, the public often realises that the qualities which make the product excel are provided by engineers, but in my own field – the design of buildings – the engineer's role is less understood and therefore undervalued. Nearly everyone in Britain, and perhaps elsewhere too, thinks that creative design in building is exclusively due to architects.

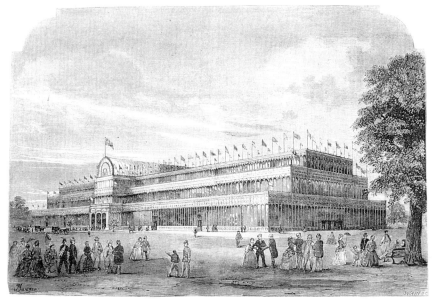

L. TO R: RICHARD ROGERS AND RENZO PIANO, CENTRE POMPIDOU; CRYSTAL PALACE AT TIME OF 1851 GREAT EXHIBITION

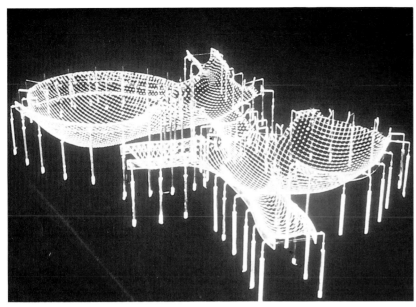

L TO R: OVE ARUP AND PTNRS/FREI OTTO. TENSILE NET FORM; COMPRESSION SHELL BUILT WITH TIMBER LATHES

But in building, the product is usually a complex one, requiring many skills in order to put many values into it. I am a building or structural engineer working in partnership with architects and others, each group bringing a body of knowledge, experience and sensibility to a common problem.

Today construction is about big money and to handle that successfully demands toughness and rigour. Autocracy or selfishness are not called for, but a system of collective decision making is essential. For such a partnership means mutual authority and shared recognition amongst the members of the building team.

I hope this magazine and the exhibition it springs from will go some way towards improving public perception of the development of art in technology and how it can relate.

Engineers, almost by their nature, excel at group work and avoid extravagant claims. They are very conscious that design usually requires many specialists who are designers in their own right and who put different qualities into the product. Engineers are sensitive of claiming sole authorship.

I am a structural engineer with an interest in building physics. Structural design is primarily concerned with the choice of form; the forces on that form and the analysis of its behaviour follow on from that choice. The whole process is influenced by the need for feasibility of execution, as success is proved in use.

Which qualities are essential to good structural engineering? Engineers should have an interest in the behaviour of materials and a knowledge of the physics of the environment. We need to give value for money. As Herbert Hoover said: 'An engineer is a man who can do for one dollar what any fool can do for two'. And I think our ambition is to achieve elegance as well as value; 'elegance' in the mathematical sense, meaning economy as well as appropriateness. Appropriateness (or function) + economy = value. As a French aircraft designer once said, 'When you cannot remove any element then you have the right design'. And here of course we can learn from nature in which structures have to be totally appropriate and mistakes become extinct.

The need for better living conditions for more people makes the engineer's struggle for efficiency worthwhile. I am of the generation which, largely due to accelerated technological changes in construction during World War II, came into engineering because of an interest in the efficient use of materials. This interest has pervaded my work as well as that of my

L TO R: OVE ARUP AND PTNRS/FREI OTTO. DIAGONAL CABLES STIFFENING SHELL; SHELL WITH COATED FABRIC

EDMUND HAPPOLD

L. TO R: PRECEDENT FOR FORCE CARRIED IN TENSION, BROOKLYN BRIDGE; ROOF REINFORCED WITH STEEL CABLE

colleagues and many others of my generation. We join a long line of engineers who have been working at long span, large space enclosures. The concrete shells of Nervi and Candela were modern versions of historic solutions; products of local materials and skills plus the advantages of mass in hot climates. Such structures are still relevant and we continue research into their construction methods. Buckminster Fuller emulated the Victorian engineers by following traditional forms, but he also copied nature in reducing materials. The Pompidou Centre was not intended to be, but ended up as, a pastiche of the Crystal Palace – and what an expression of intermediate space that was. The high energy input/output ratio of steel has led some to become interested in timber. Thus the German architect, Frei Otto, using an equal mesh, three-dimensional version of Robert Hooke's hanging chain model has defined a tensile net form which, when reversed, provides a pure compression shell; and proposed that it be built with timber lathes. It would have collapsed. To act as a shell it needed diagonal cables to provide sheer stiffness but, because of its lightness, it was very economic to erect. It was test loaded with the town's dustbins and covered with a coated fabric.[7]

To carry a force in tensions is, of course, in material terms the most effective building solution and a very old one. Reinforced with steel cable, such roofs can safely and economically cover several acres. The actual skin can be concrete – as in Calgary, designed by Jan Bobrowski with our partnership as proof engineers – or it can be PVC, PTFE-coated fibre glass, timber shingles, ceramic tiles or even, nowadays, stained glass. But most loading is from the wind and snow and thus varies with the season. Steel cannot compare with the energy-storing characteristics of timber – but, if the steel wires are crimped to act in the more lightly loaded condition like a spring, it can store energy mechanically. We have used crimped steel mesh as a fabric for aviaries in Munich, San Diego and Hong Kong. Compare the similarity of the mesh to a spider's web.

But, to return again to timber, recent comparative studies have shown us that it is the strain energy characteristic which also reduces the cost so radically. Timber, in proportion to its weight, is comparable in strength and stiffness to high strength steels. But the problems with timber have always been in achieving an effective tension connection between members, and it is only since the discovery of epoxy adhesives, impregna-

L. TO R: BURO HAPPOLD/JAN BOBROWSKI, CONCRETE SKIN OF CALGARY; STEEL WIRE AS ENERGY STORE

L TO R: METAL ROOF COATING; BÜRO HAPPOLD/FREI OTTO, DIPLOMATIC CLUB, RIYADH, SAUDI ARABIA, STAINED GLASS ROOF COATING

ted with fibres developed for windmill blades in the USA, that 'high strength' collinear connections have been made possible. We can now achieve 85 percent full strength.

For many years carpenters have resisted removing any more of a tree than the bark. This is because the tree, to protect its outer capillaries from buckling under sudden gusts of wind, orientates some of its fibres diagonally around the capillaries so that an element of longitudinal pre-stressing is given to the outer fibres containing the sap (to the order of 14 MN/m^2), while the centre of the tree is in compression. When the wind blows, the outer capillaries are well able to carry the tension.[8] Combining advances in timber connections with the tensile properties of raw timber has made possible the hanging forms in the School for New Woodland Industries at Hooke Park on which we have worked.

It becomes interesting to use structure not only to resist forces but also actively to amend the internal environment. Light, of course, is the most powerful source of energy, and designing covered cities in the Arctic is simply an extreme instance of this. In order to achieve a satisfactory, all-year-round environment under such a cover, the quality of light is all important and that quality is dependent on as much of the spectrum as possible being transmitted. This is why glass is used for windows, even though it does reduce the ultra-violet part of the spectrum, thereby creating a 'greenhouse' effect. Studies have been carried out for an air-supported cover over a proposed 36-acre city in the Athabasca region of Alberta. These show how some of the new laminates transmit more of the light spectrum than glass – even making possible grass cultivation in the Arctic conditions. Alas, the collapse of the oil market caused the project to be dropped. This development has moved slowly since then, but one of the German manufacturers has used this laminate as a cover in building a conservatory for tropical birds and alligators at Arnhem in Holland, and we used the lighting aspects of the concept for a roof in Brunei. It seems to perform very well.

Of course, our bubble idea is still relatively crude and needs further work. But one cannot help but be interested in further developing the possibilities of utilising energy from light. After all, this is only the old idea of the conservatory or greenhouse. When it is too hot we shade ourselves, like plants do with their leaves. But we still cannot build firm organic substances from carbon dioxide, water and light, as plants do. We must find out

L TO R: PVC ROOF COATING; CRIMPED STEEL MESH FABRIC; TIMBER SHINGLE ROOF COATING

L TO R: CRIMPED STEEL MESH FABRIC FOR MUNICH AVIARY; SPIDER'S WEB AS WIRE PARADIGM

how, before the slight surplus store of carbon – coal, oil, gas and timber – which has been built up over millions of years and which at the moment provides most of our structural materials, is finally depleted.

Engineers are moving towards it. In fact organic materials are becoming more and more important. And just as 200 years ago those who resisted change thought iron and steel 'inhuman' materials, today people say the same about plastics. Yet such materials, in composites and laminates, are essential materials in aerospace and their use is growing.

All this is about change. Yet much engineering is still at a simple level. The pace of advance in technology is generally set not by the most brilliant and able engineers but by the capacity of the individual – engineer or skilled mechanic – to master and use an improvement efficiently and harmoniously. My last example illustrates this: an exposed bridge over a busy road in a town. A simple steel trussed solution can be prefabricated and erected in a matter of hours. Then it can be clad like a building to reduce exposure and deflect the wind. It becomes a simple piece of street furniture.

You may have noticed that most of the ideas in these projects

were first developed for overseas clients. Why is new technology not better utilised in Great Britain? The Prince of Wales is promoting community architecture in the cities. Yet there are few problems of the inner cities which could not be solved by the economic well-being which productive industry can provide. It is engineers who could bring this about. Our car industry is a failure, yet we have engineers who design the fastest, safest and most reliable cars in the world. Our car body stylists, trained at the Royal College of Art, are in demand in Germany, the United States and Japan, where they work in Mercedes Benz and other successful car firms. If only these two groups could work together in Britain as equals, they could succeed.

Yet our intellectual and economic class systems are such that this seems impossible. We have the skills but not the business or social structure to enable our engineering expertise to revitalise our industry. Perhaps the failure is managerial. What is common to most successful overseas technical enterprises is the inevitability of collective decision making. Could it be that our national belief that someone should be in charge – preferably on an annual financial basis – inhibits this? Maybe those Quaker industrialists with their collective long-term ambitions were

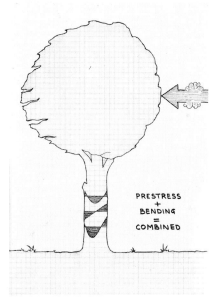

PRESTRESS
+
BENDING
=
COMBINED

NO ONE CAN accuse the designers of the building shown here for failing to see the wood for the trees. They have come up with a method of extracting timber for house construction which, far from denuding Britain's forests, actually leaves them flourishing.

Green roundwood technology is a combination of primitive woodcraft, avant-garde engineering and a dash of high-tech materials which uses timber in its unsawn state.

Britain's conifer plantations cannot produce the high quality timber needed for conventional methods; but they do produce 3m tonnes annually of thinnings, low-grade timber which is literally weeded out to make room for the rest. The thinnings are usually pulped or burnt after.

Professor Ted Happold, head of the school of environmental

Logs of ideas

design at Bath University, argues that timber is stronger in its natural round state. He says its fibrous structure makes it particularly strong in tension, although it is usually used only in bending or compression.

Happold and John Makepeace, director of the school of timber design at Hooke Park, Dorset, have constructed a prototype building using struts, ties and cables instead of conventional posts and beams.

The main problem of the new approach is the joints. Epoxy resin is often used, but it is slowly degraded by the ultra-violet component of daylight. Happold came up with the idea

of gouging out the ends of logs so that the glue would be shielded from the daylight.

The wood's longitudinal cell structure can be exposed so that the epoxy resin is drawn deep inside the log. If natural wood fibre is added to the epoxy resin, the glue can be made to resemble the timber. The fact that the timber is still moist with sap actually works to the joints favour, as it permeates both timber and glue.

According to Makepeace, green roundwood offers Britain a new indigenous constructional resource that avoids the wholesale destruction of forests, and which could provide up to 30,000 houses a year.

Martin Spring

L TO R: CRIMPED STEEL MESH FABRIC FOR HONG KONG AVIARY; TREE FIBRE STRESS; HANGING TIMBER FORMS

L TO R: AIR-SUPPORTED COVER FOR PROPOSED 36-ACRE CITY ALBERTA; LAMINATED ROOF COVER, ALBERTA

right and we should consider returning to those lost managerial values.

Notes

1 J K Galbraith, *A Life in Our Times*, Andre Deutsch, 1981

2 I Raistrick, *et al*, *Dynasty of Iron Founders*, David & Charles, 1970

3 A W Skempton (ed), *John Smeaton FRS,* Thomas Telford London, 1981

4 Persig, *Zen and the Art of Motor Cycle Maintenance*, Corgi, 1976

5 Martin J Weiner, *English Culture and the Decline of the Industrial Spirit, 1850-1980*, Cambridge, 1981

6 Roger Fry, *Vision and Design*, Chatto & Windus, 1920

7 E Happold and I Liddell, 'Lattice Shell for Mannheim', in *Structural Engineer*, 1975

8 James Gordon, *The New Science of Strong Materials*, and *Structures*, both published by Penguin, are two books which cover these subjects marvellously. I owe the author an immense debt for the stimulus he has given me, both personally and in his books

Professor Edmund Happold is Head of Engineering and Architecture at the University of Bath, and Senior Partner of the engineering practice, Büro Happold.

L TO R: LAMINATED ROOF COVER, ARNHEM, HOLLAND; PROPOSED 'STREET FURNITURE, PREFABRICATED BRIDGE'

ROBERT THORNE

WAITING FOR THE QUEEN (LITHOGRAPH AFTER JOSEPH NASH)

ROBERT THORNE
Paxton and Prefabrication

ENGRAVING OF CRYSTAL PALACE FROM THE *ILLUSTRATED LONDON NEWS*

In the history of prefabrication the Crystal Palace has always been awarded legendary status. The story of its design and erection, which never seems to wear thin however often repeated, fully justifies that reputation, but even the most compulsive accounts of the building works in Hyde Park and later at Sydenham cannot conceal that it is a story with a slightly unsatisfactory ending. Within a matter of years, if not months, an ideal method of component construction was

developed to a peak of efficiency in preparation for the most celebrated public event of the 19th century; and then almost as fast it faded from view, confined to marginal projects none of which matched the success of the prototype. Descriptions of the Crystal Palace generally reach a crescendo at the opening of the Great Exhibition, with a less hearty burst of enthusiasm at the opening of the Sydenham building, and then leave the rest to silence. The hiatus which followed, which is in its way just as instructive to any discussion of prefabrication, has gone largely unconsidered, perhaps above all because lessons based on non-events can never hope to match those that deal with things that really happened. But it is also the case that the failure to examine the aftermath of the erection of the Crystal Palace has helped keep from view some of the more problematic aspects of the building's history.

The conventional account of the preparations for the Great Exhibition is by now so familiar as to hardly need reiteration.[1] It assigns an heroic role to Sir Joseph Paxton, once an humble gardener's boy, who saved the Commissioners of the exhibition from the apparently impossible quandary of how to house the event. In alliance with the contractors, Fox Henderson, he was able to exploit a loop-hole in the document inviting tenders for the Commissioners' own design so as to secure acceptance of his alternative scheme, while at the same time he published his design in order to rally public support behind it. The same publicity machine which helped win victory for his proposal stayed in action to celebrate its miraculous realisation in Hyde Park during the winter months of 1850-1. Led by the *Illustrated*

London News, press reports of what was happening behind the site hoardings provided a more detailed coverage of the construction process than any other building has been privileged to receive: the raising of the columns and girders, the cutting of sash-bars, the manufacture and fitting of the glazing, the debates about paint colours – nothing was thought too esoteric or technical for inclusion. Through the accumulated effect of such reports the building was acclaimed for the speed and ingenuity of its creation well before any reasonable assessment could be made of its final effect.

In a sense the Crystal Palace was a gift to the press because its process of assembly was so easy to grasp. The wet and messy business of conventional masonry construction, which left the spectator at a loss to understand the progress and importance of what was happening, was supplanted by an easily understood sequence of construction. In design, the key to its modular system was in its roof. The largest feasible panes of glass for use in Paxton's ridge-and-furrow roofing system were 49 by 10 inches. Two such panes set at the correct angle constituted a roofing segment 8 feet wide: from that dimension stemmed the standard lengths of roof and gallery girders (24, 48, and 72 feet) and thus the dimensions of the building as a whole. Strictly speaking, therefore, the entire building, 1,848 feet in length with its galleries, aisles and transept, obeyed a structural logic based on just one of its features.

As with any prefabricated structure, the principal virtue of this standardisation was in permitting most of the work to be done off-site, in this case at the Fox Henderson factory in

Smethwick and at other firms in the West Midlands and London. But much as that helped speed the process, it depended on sophisticated timing to organise the arrival of components in the right order for erection. The chief attraction of the site in Hyde Park while work was in progress was the spectacle of the Fox Henderson team orchestrating the receipt of the different parts, testing them where necessary, and then dispatching them to the right spot for assembly. With the system working at full tilt, castings could be ready for use in Hyde Park within 24 hours of leaving Smethwick foundry, and Paxton claimed to have seen two columns and three girders erected in just 16 minutes.

As if the story of its erection was not enough, the Crystal Palace could claim to be a landmark of prefabrication in two other senses. First, its removal from central London and re-erection in an adapted form at Sydenham demonstrated that a building made up from a kit of industrial parts could be dismantled as easily as it had been put up, with the capability of being reused in a variety of other forms. In that respect it had more in common with the machines which helped bring it into being than with buildings of traditional construction. And secondly, an account of its design was published that was sufficiently exact for

course of proceeding that would not be recommended for imitation, and by omission of what would be in every other case an architect'. And whenever the subject of iron came up in debates at the RIBA the example of the Crystal Palace was taken as testimony that the brutish qualities of such a material needed a proper architectural sensibility to tame and develop them. 'In my opinion', announced the sententious Robert Kerr, 'the more the system of iron and glass is amplified, the more it will be seen that no really permanent scientific work can be produced by that means; and, though some excellent effects may be accomplished for a moment, yet architecture at large is not so greatly advanced constructively, as we might expect by that innovation.'[3]

The territorial wranglings in the building world that helped engender such remarks may also be understood as part of a wider reaction in which the reputation of the Crystal Palace was bound to suffer. Architectural historicism of the kind which Kerr and his contemporaries took as their starting-point was one manifestation of the rejection of utilitarian rationalism which took hold in the 1850s. The Crystal Palace was caught up in that reversal because it was a symbol of a way of thought that was

DELAMOTTE. CONTEMPORARY VIEWS OF THE DISMANTLING AND RE-ERECTING OF CRYSTAL PALACE, 1852

it to be a template for other buildings following the same system. The prime intention of Charles Downes and Charles Cowper in issuing their volume of the Fox Henderson working drawings was to round off the technical record of the Hyde Park building, but as they were quick to point out, the book which resulted was 'a work so correct and complete as to enable an architect or engineer to erect a similar building if necessary'.[2]

So at first glance the Crystal Palace seems to have fulfilled more than a fair share of the requirements expected of a prefabricated building – in its mode of construction, its adaptability, and its potential as the basis for a repeatable architectural system. In the opinion of many 20th-century commentators its catalogue of virtues was such that any discussion of why its promised progeny failed to appear is bound to look to causes beyond its immediate sphere. Among such explanations the most readily cited has been the professional jealousy that the Hyde Park project engendered; the fact that architects (and consulting engineers for that matter) were pushed aside by an outsider in a project of such significance continued to rankle for many years after. *The Builder*, in its obituary of Paxton, emphasised that 'the building of 1851 was accomplished only by a

running out of favour and it suited many people to single it out as a victim.

Such explanations, pragmatic or philosophical, for the failure of the Crystal Palace to initiate a general architectural transformation, presuppose that the building itself was almost faultless and that the methods which brought it into being could have readily been applied elsewhere. In the 1930s, when the building was canonised as a classic of modular construction, its lessons seemed so urgently obvious that the reluctance of previous generations to heed them could only be attributed to jealousy and short-sightedness. 'To my knowledge', said Siegfried Giedion, 'the possibilities dormant in the modern civilisation we have created have never been so clearly expressed'.[4] All that was required was to pick up where Paxton left off.

Half-a-century later, the projects of those who were nurtured on that view of the 19th century have their own tale to tell, and in the knowledge of what they achieved the building which they looked to as an exemplar takes on a different hue. In the emergencies of the post-war period a Paxtonian solution was more than simply a way of closing an historical gap: for architects designing schools and housing in the public sector it seemed the only way to meet the building crisis. By background and necessity they were attracted to industrialised building techniques; to the Hills system, the Derwent timber frame, Intergrid, and most famously, to CLASP. How these systems were developed, in an almost Victorian atmosphere of energy and expediency, has been told with proper sympathy by Andrew Saint in his book, *Towards a Social Architecture* (1987). As with so many episodes it turns out to have been a more complicated and uneven matter than is generally supposed. By abstracting the lessons of his account, and applying them to what happened at the time of the Great Exhibition that story in its turn loses some of its heroic simplicity.

As post-war architects readily appreciated, putting buildings on the production line requires a different approach to every aspect of the building process. For a start, as Paxton demonstrated when he delivered his preliminary drawings to Fox Henderson, the component manufacturer is bound to have thrust upon him many aspects of design, analysis and research which traditionally are the architect's prerogative. The question of what is feasible in conditions of industrial production is as critical as conventional aspects of design, so responsibility for

be discussed primarily in terms of his career. The Crystal Palace was a greenhouse writ large, and his previous experience was in the design of garden structures, so the links can easily be made. In its most horticultural qualities it was indeed a metropolitan version of what he had accomplished at Chatsworth, particularly in its glazing, its ridge-and-furrow roofing (a system devised to improve plant propagation) and its use of standard, machine-cut wooden parts.[6] But nothing in Paxton's previous career had prepared him for a project of such a scale, nor for one requiring the extensive use of cast and wrought iron, materials which he never felt at home with. Hence his dependence on the firm of Fox Henderson, not just for submitting a tender on his behalf but for taking the risk to, as Sir Charles Fox put it, 'mature and realise' his scheme.[7]

The Fox Henderson contribution, though less romantic, puts the Crystal Palace in a more realistic context. The company had started life in 1839 as Bramah, Fox and Company, a partnership of John Joseph Bramah and Charles Fox. Its works in Smethwick, designed by Fox, was completed in 1841, in time to catch the engineering contracts of the railway boom in the mid 1840s. By 1845 Bramah had retired, and the firm changed its name to

L TO R: DISMANTLING AND RE-ERECTING CRYSTAL PALACE AT SYDENHAM: INTERIOR AT TIME OF EXHIBITION

shaping a project inevitably shifts towards the manufacturer, although whether credit for that responsibility shifts as well is another matter. At the other end of the process, prefabrication presents another kind of challenge to customary architectural practice. By definition, a modular system, however flexible, is not tailored to a single site and can (as critics of post-war schools never tired of pointing out) seem clumsy in many locations. If it works well, and stands the test of other expectations, that may be counted only a slight drawback, but to many eyes the individuality of a building in a particular setting is a first consideration, not an added luxury. As Andrew Saint concedes in his assessment of system-built schools: 'There is no getting away from appearances and aesthetics'.[5] If that is so for buildings of one type, most of them modest and unobtrusive, it becomes doubly problematic when the application of a prefabricated system to a variety of buildings is proposed.

Returning to the Crystal Palace, how do the lessons of subsequent prefabricated projects alter the popularly accepted view of the role played by the manufacturer? As long as Paxton is cast as a Napoleonic figure armed with a visionary scheme and the ability to see it realised, the ancestry of the building is bound to

Fox Henderson and Company with the arrival of the Scottish engineer, John Henderson. As far as is known, Fox was more of a designer while Henderson looked after running the works. Though the company fabricated iron products of every description (including prefabricated houses for California) its attraction for Paxton was its reputation for the design and manufacture of iron roofs. Two of its station roofs have long been known, though the structures themselves have been demolished: Tithebarn Street, Liverpool (1849-50), designed in conjunction with John Hawkshaw, and New Street, Birmingham (completed 1854), designed by one of the firm's staff at Smethwick.[8] Other roofs, constructed to cover ship-building slips in naval dockyards, have recently won comparable attention, thanks to the detective work of James Sutherland. Fox Henderson's first naval contract seems to have been for two roofs at Pembroke Dock, begun in 1844, both of them composite trusses spanning 80ft 8in, with 20ft overhanging eaves.[9] They have not survived, but a similar roof of about the same date was erected by the nompany at Woolwich and subsequently, after that dockyard closed, reappeared in a transmogrified state as No. Eight Machine Shop at Chatham, where it can still be seen. Nearby at Chatham is another migrant Fox Henderson roof: the plaque on the Boiler Shop, dated 1876, marks its re-erection there after a previous life as the roof of No. Four Slip at Woolwich.

True to Paxton's predilections, the Hyde Park Crystal Palace had a great deal of wood in it, but, because of the iron structure on which it depended, Fox Henderson were a natural choice as contractors. Their role provides a more convincing connection between the project (plus the Exhibition it housed) and the industrial society which spawned it than an ancestry derived from hothouses and gardening. Their commitment made them, as was remarked at the time, 'masters of the situation', from the first stage of preparing working drawings and making structural calculations, through the fabrication of components and their assembly on site, to the eventual dismantling of the structure and its transfer elsewhere.[10] On the apparent evidence of such a chronicle of success the project seems to vindicate the prominent role which prefabrication gives to component manufacturers. Yet, as post-war experience has demonstrated, such involvement can also be a costly trap for a company which is lured into taking a greater responsibility for design and performance

that claim, or the professional interest which fuelled it, the transept certainly was the idol of those who doubted whether a modular structure, unrelieved by such a feature, could achieve a pleasing effect. As a contributor to the *Architectural Quarterly Review* put it: 'The interior of the transept of the Exhibition Building, with its arched roof – due to Mr Barry – has ... a noble effect; but this stamps only more completely the inferiority of ART in the general building'.[12] None doubted that the perspective down the nave, heightened by Owen Jones' clever colour scheme, had a dazzling impact on first viewing: 'fairy-like' were the words on everyone's lips. But translucent length on its own was not enough. Apart from the transept, nothing else interrupted the rhythmic monotony of columns and girders, which in repetition were eventually confusing and tiring.

As for the exterior, Henry Mayhew described the 'disappointment on the countenances of the newcomers on first beholding the building', as they realised how far the magazine illustrations were from reality. Except from certain angles, the magnificent scale of the building was hard to grasp:

Look along it, and the trees soon block out the view; look up at it from either side, and the extreme width of the

DELAMOTTE, CONTEMPORARY VIEWS OF CRYSTAL PALACE INTERIORS

than it had allowed for. Fox Henderson, despite their track record on previous engineering projects, do not seem to have wholly escaped such risks. In November 1851, after the Exhibiton had closed, the firm was granted an additional £35,000 on top of its original contract of £79,800 plus later additions. This was necessary, said the Commissioners, because 'Fox Henderson discovered that their net liabilities were far beyond those to which they were entitled in the contract'.[11] Even taking account of the fact that the contract was a complicated one in the first place, the Commissioners' generosity, in the flush of the Exhibition's success, conceals a significant drawback in the method used to get the event housed on time. In less sympathetic circumstances Fox Henderson might never have seen the year out.

In the other principal debate which prefabrication has engendered the question whether an industrial method can produce an acceptable architectural aesthetic – the Crystal Palace again has a less secure position as the indisputable ideal of its kind. Some of the resentment that the architects harboured towards Paxton derived from the belief that one of their number, Sir Charles Barry, had suggested the round-arched transept, the most highly appreciated feature of the building. Regardless of the truth of

building prevents you seeing the upper storey, so that it has a most unseemly 'squat' or 'dumpy' appearance.

And approaching closer, the building belied its name by presenting a ground floor sheathed in wooden hoardings.[13]

If the critics of the Crystal Palace had been simply a few disparate voices – disgruntled architects, militant Ruskinians, and those who find something to complain about in everything – their opinions might reasonably be discounted in any final judgement of the project. What gives weight to their views is that those most responsible for the building recognised their validity and endeavoured, in the years immediately after the Exhibition closed, to take account of them. Paxton, Fox Henderson and other members of the building team went on, singly or together, to other projects which were regarded as the direct successors to the Hyde Park one, notably the re-erection of the Crystal Palace at Sydenham and the construction of the new GWR terminus at Paddington. As they did so they fell over themselves to show that they too were eager to advance iron and glass architecture beyond the 'scaffold pattern' of the prototype.

Paxton had always hoped that the success of the Crystal Palace would ensure its retention in Hyde Park as a winter garden in which 'the great truths of Nature and Art would be constantly exemplified'.[14] But having insisted in almost the same breath that it could easily be dismantled, his campaign to keep it standing had a difficult case to make: defeat was conceded with the government's decision in March 1852 that the park should be cleared. Two months later the Crystal Palace Company, established to organise the building's resurrection as a temple of national recreation, had issued its first prospectus: by August the same year the first column had been raised at Sydenham. Paxton was the moving force in the company; Fox Henderson were again the contractors, with John Cochrane and John Henderson in charge; and many others associated with the Great Exhibition had a part in fitting up the interior.

The progress of reconstruction never attracted the same attention as the original project, except on the melancholy occasion in 1853 when a portion of the timber scaffold used to erect the ribs of the central transept collapsed, killing 13 men. But even relying on just the inquest reports from that event it is easy to appreciate that the Sydenham structure was much more than simply a re-run of its predecessor.[15] The ribs of the celebrated Hyde Park transept were of laminated timber, assembled

the redesign were not as prone to make public statements about their intentions as they had been first time round, but the effect of what they did was to produce a tighter, more architectural treatment, creating as much of an effect of permanence as its prefabricated structure allowed. The next possible concession to critics might have been to produce a yet more articulated profile by adding the three domes that Barry had suggested, one at each of the transept intersections. But even that might not have been enough. Barry's son, speaking in 1871, kept up the chorus that his father's friends had started:

I cannot conceive that anybody can look upon that structure as either artistic or of scientific construction. It is a huge glass case, situated in a position where the eye wanders from the building to the beautiful surroundings of Nature; therefore it gets a great amount of adventitious praise of which, architecturally, it is undeserving.[17]

Paddington Station, completed the same year as the Sydenham Crystal Palace, offers a yet more instructive example of how the lessons of the Exhibition building were absorbed almost before they had happened. I K Brunel, hurt that his own proposal for accommodating the Exhibition had been so mauled

CONTEMPORARY VIEWS OF CRYSTAL PALACE INTERIOR AT TIME OF GREAT EXHIBITION

on the ground before being raised into position in pairs. The new central transept was higher, wider (120 as against 72 feet) and built of wrought iron; its principle ribs, set in pairs at 72 foot intervals, were of uniform depth with their top and bottom flanges braced by a lattice of flat diagonal bars. Each pair of ribs was carried, via connecting frames, on four columns, two of which were set forward from the main line of the gallery. The scaffold collapse related to the problem of fabricating a roof for which the Hyde Park building provided no precedent. On the same count, although the transept (like the rest of the building) incorporated columns and girders which had already done one term of service, it also called for hundreds of new castings fresh from the Fox Henderson foundry.

Changes of the kind to be seen in the central transept were repeated in almost every other aspect of the rebuilt structure. At Sydenham two end transepts were added, built on the same principle as their larger brother, and the nave was raised to their status by being given an arched roof. This remodelling took place within a length 10 bays less than the original, and appeared yet shorter because the Jones colour scheme had disappeared beneath dignified coats of dark red. The team responsible for

by other members of the Building Committee, was ready to demonstrate the virtues of what he dubbed the 'railway shed style'. Early in 1851 he enlisted Matthew Digby Wyatt, known to have the same views, and Fox Henderson, who were than so delirious with the progress of their Hyde Park contract that they felt they could tackle anything. Independent minded though Brunel was, he and his team could not avoid conceiving of the station as anything other than a mature variant of the Crystal Palace.

Today, looking along the lines of columns which support the triple-arched roof, the parentage is unmistakable, even though their spacing is wider than in the Hyde Park building and the column to girder connections are differently handled. Also, the sheds are twice interrupted by 50 foot wide transepts, introduced to modulate and relieve their length just like the one which Paxton had been persuaded to adopt. But a description of the station cannot be founded on such analogies alone, leaving unremarked the parts of its structure which stretch the Paxtonian model into a more consciously architectural form. The wrought iron arched ribs used at Paddington had no precedent in the original Crystal Palace, nor had the idea of piercing such ironwork with a pattern of stars and planets to relieve its tunnel-like effect. In other kinds of decoration, as well, Brunel and Digby Wyatt were innovators: the swirling iron tracery of the end gables, the column capitals, and the foliage pattern which sprouts in the haunches of the ribs. 'A station after my own fancy', as Brunel called it, was not just a workable terminus but a lesson in how the temporary qualities of iron and glass could be reinterpreted for enduring use.[18]

Paddington received high praise ('by far the finest work of its kind in Europe', pronounced *The Builder*) but on its own was not enough.[19] The new order of architecture, as dreamt of by Paxton, had to extend further than obvious kin such as railway stations to make his case proven. He was ready to talk in terms of housing, churches and factories, not to mention his old stamping ground of horticultural buildings, but in practice his movement (if it can be so called) ended where it had started, with exhibition buildings. Such structures, by their very nature exotic and ephemeral, were unhelpful in establishing a continuity of experience in modular construction, though they might advance debate and practice in certain ways. But as it turned out, those responsible were so conscious of the assumed faults of the Crystal Palace that their contribution was to rein in the tendencies which it had started, without providing a strong alternative. Owen Jones, an old Crystal Palace associate, was almost alone in advancing the cause of iron and glass for exhibitions, but none of his designs were built.

Of the exhibition buildings that were completed in the 1850s, the New York Crystal Palace of 1853 shows the often bizarre way that the lessons of Hyde Park, valid and otherwise, were digested. An iron and glass building was stipulated by the New York City Council and of such designs submitted in competition (including an entry by Paxton himself) one by the partnership of Carstensen and Gildermeister was chosen: a Greek cross surmounted by a dome, with lean-tos marked by towers in the angles of the cross. 'Columns, girders, and arches', it was said, 'will be connected according to the system of Messrs Paxton, and Fox and Henderson'. Whatever form of borrowing that constituted, 'the monotony of Mr Paxton's design' was eschewed in the adoption of a plan and system of embellishment which consciously suppressed memories of the Hyde Park model. This was a project completed before the lessons of the Sydenham Crystal Palace, or indeed Brunel's Paddington, were available, but it illustrates a movement which their influence could do nothing to deflect.[20]

If any one event marked the termination of the expectations first engendered by the Crystal Palace it was not the death of Paxton in 1865 but the collapse of Fox Henderson nine years before. The immediate cause of the failure was a loss of about £70,000 on a Danish railway contract but the indications were clear, from their abysmally slow completion of their two Paddington contracts if nothing else, that the reputation they had gained at Hyde Park had got the better of them.[21] Even there, as already suggested, their conjunction of skills was not as wholeheartedly satisfactory as is generally thought. Other firms took their place, but no other engineering contractors applied their own skills to the cause of structural innovation to quite the same degree as Fox Henderson; and perhaps for that reason none could claim a roll-call of projects to compare with those that passed through the Smethwick works in the decade 1844-54. But even if Fox Henderson had survived, that would not have saved the day for prefabrication. They brought into being a model system and helped mature it, but nothing they could do would have ensured its perpetuation.

Notes

1 Christopher Hobhouse, *1851 and the Crystal Palace*, 1937; Yvonne French, *The Great Exhibition: 1851*, 1950; Ralph Lieberman, 'The Crystal Palace', *AA Files*, 12 Summer 1986, pp 46-58

2 Charles Downes and Charles Cowper, *The Building Erected in Hyde Park for the Great Exhibition of the Works of Industry of All Nations*, 1852, reprinted 1971

3 *Builder*, June 24 1865, p 443; *RIBA Transactions*, 1st Series, Vol XXV, 1874-5, p 214

4 Quoted in *Architectural Review*, Vol LXXXI, Feb 1937, p 66

5 *Towards a Social Architecture*, 1987, p 235

6 On Paxton at Chatsworth see George F Chadwick, *The Works of Sir Joseph Paxton 1803-1865*, 1961, pp 72-103, and articles by Chadwick on the Great Stove in *Architectural History*, Vol IV, 1961, pp 77-91, and Vol VI, 1963, pp 106-9

7 *Illustrated London News*, July 5 1851, p 21

8 'Description of a Large Roof Recently Erected at the Liverpool Terminus of the Lancashire and Yorkshire Railway', *Transactions of the Royal Scottish Society of Arts*, Vol IV, 1856, pp 94-7; Edwin A. Cowper, 'Description of the Wrought-Iron Roof over the Central Railway Station at Birmingham', *Proceedings of the Institution of Mechanical Engineers*, 1854, pp 79-87

9 Capt M Williams, 'Description of Wrought Iron Roofs Erected Over Two Building Slips in the Royal Dockyard at Pembroke, South Wales', *Papers of the Corps of Royal Engineers*, Vol IX, 1847, pp 50-58

10 'Helix', (pseud, W Bridges Adams), 'The Industrial Exhibiton', *Quarterly Review*, Vol LV, July 1851, p 354

11 First Report of the Commissioners for the Exhibition of 1851, *Parliamentary Papers*, 1852, XXVI, p 30

12 *Architectural Quarterly Review*, June 1851, Vol 1, p 26

13 *Edinburgh News*, May 3 1851, p 4. For similar remarks by a visiting Dutch architect see *Builder*, Oct 18 1851, p 655

14 Joseph Paxton, *What is to Become of the Crystal Palace?*, 1851, p 13

15 *Builder*, Aug 20 1853, pp 529-30, Sep 17 1853, pp 590-1, Sep 24 1853, pp 602-4; *Civil Engineer and Architect's Journal*, Sep 1853, Vol XVI, p 355-8 The collapse of the scaffold seems to have been caused by the failure of the upper part of a truss (principally a jointed beam 90 foot long) while it was being moved into position.

16 C R von Wessely, 'On Arched Roofs', *Civil Engineer and Architect's Journal*, Apr 1866, Vol XXIX, pp 107-8

17 RIBA Proceedings, 1871-2, p 84

18 Robert Thorne, 'Masters of Building: Paddington Station', *Architects' Journal*, Nov 13 1985, pp 44-58

19 *Builder*, Dec 24 1870, p 1020

20 George Carstensen and Charles Goldemeister, *New York Crystal Illustrated Description of the Building*, NY 1854, p 48

21 *Illustrated London News*, Nov 1 1856, p 441; Thorne, *op cit*, p 51

Dr Robert Thorne is a member of English Heritage.

ROBERT THORNE

FRANCIS PUGH
Design, Engineering and the Art of Drawing

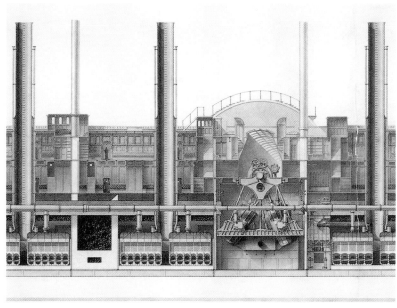

I K BRUNEL, SS *GREAT EASTERN*, SECTION, *c*1857-60. (DRAWING BY SCOTT RUSSELL)

A sophisticated computer graphics programme can produce multiple views of an object rapidly and cheaply, a task which not so long ago required the time-consuming labour of a large drawing office. Yet the sheer facility of this process can easily make us forget the conceptual power of the basic language of drawing. One has only to imagine the process of design, manufacture or construction without such a system to realise that in common with other tools of prediction and

analysis, it greatly expands our capacity for sustained innovation. But how did this language evolve and what were the circumstances that brought its various elements into being?

Architectural Drawing

Even in the distant past any major work which required the deployment of large numbers of craftsmen, also needed a means for organising and directing their labour. Plans drawn by master masons served this purpose in medieval Europe and had done so in some form since Graeco-Roman times, but for the most part traditional methods of full-scale drawing were sufficient for even the largest undertaking. The shapes of vaults and arches were inscribed on the plaster floors of mould lofts, while stone-masons and carpenters used templates as well as marking directly onto materials. Drawings of elevations seem first to have appeared towards the end of the Middle Ages when the complexity of surface design had reached a point where the work of the stone carvers needed more precise direction. But to our eyes medieval drawings lack a coherent system for relating the parts of a structure to the whole. The system which seems so familiar today is based on the concept of scientific perspective, a discovery usually attributed to the Florentine sculptor and architect Brunelleschi. To Brunelleschi, and later 15th-century humanist philosophers, perspective represented much more than a technical device: it symbolised a new objective view of man's place in a divinely ordered universe. They sought to understand this idea further through the study of Euclidean geometry, from which was gained a greater awareness of pattern, order, proportion

and relationship; principles which also underlay the desire to create a new harmonious architecture based on classical ideals.

Architects of the generation after Brunelleschi began the process of inventing a new basis for the language of drawing when their attempts to communicate novel architectural forms led them to the notion, implicit in scientific perspective, that three related views were sufficient to define any object or structure. Artists and architects including Leonardo da Vinci and Bramante contributed to this development but Raphael was probably the first to make use of plans, sections and elevations related in logical sequence and to a consistent scale. Whatever his underlying philosophical motive, the immediate cause seems to have been the need to oversee his many workshop commissions while at the same time fulfilling the terms of his appointment as architect of St Peter's after 1515. Drawings supplied in advance would allow building work to continue during his frequent absences from Rome. This business-like approach is equally evident in his proposal that the ruins of ancient Rome should be measured and drawn to give architects and scholars a more thorough understanding of the scale, appearance and construction of Roman buildings. In both instances, it is apparent that Raphael, in common with many of his contemporaries, held a comprehensive view of the function of drawing. Not only could it be used to implement a design, it could also serve as part of a business transaction, a record, a form of instruction and of objective analysis. Later generations of architects used drawing for all these purposes so that knowledge of current designs and knowledge of the craft of building were no longer regarded as

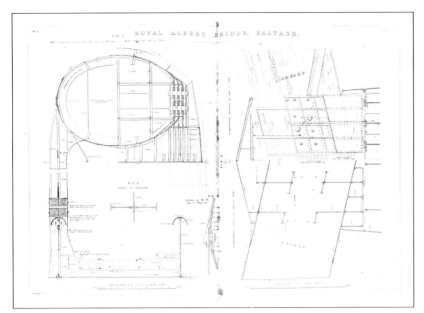

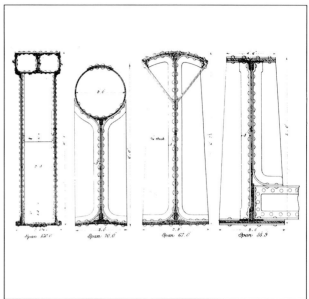

L TO R: I K BRUNEL, SALTASH BRIDGE, *c*1858, STRUCTURAL DRAWING: SECTIONS THROUGH RIVETED WROUGHT-IRON BEAMS OF 1850-60

'mysteries' communicated only to initiates in workshops and guilds, but could be freely disseminated in sketchbooks and technical treatises.

The key figure in this transition to something akin to modern practice is Andrea Palladio who began his career as an apprentice stonemason and only later became acquainted with the philosophy of humanist architecture through his patron and teacher Giangiorgio Trissino. Like Raphael, Palladio regarded measurement and drawing as more effective ways of gaining accurate knowledge about the remains of the ancient past than the study of humanist theory or the writings of Vitruvius. His own architectural treatise, the *Four Books on Architecture* of 1570, while modelled on Vitruvius, is essentially a practical work. Its wood block illustrations, though less than perfect for conveying subtleties of line and tone, were to have an influence far beyond the confines of the society and period in which they were conceived, because they conveyed, as no drawings had done before, a sense of the scale and proportion of classical architecture in a simplified and readily understandable form. Palladio's use of a lucid and orderly sequence of carefully chosen cross-sections is apparent in his drawings of the Baths of Cara-

calla. It has been suggested that this orderliness was imposed by Palladio and is not evident in the buildings themselves.[1] If so, it further emphasises the way in which our present language derives from a system of drawing that sought to organise experience for philosophical and spiritual ends while never losing sight of the practical advantages that had been gained in consequence.

Engineers and the Art of Drawing
Palladio's drawings were influential in many ways, not least in providing exemplary models for the illustrators of scientific or technical works and for the working drawings of engineers and architects. It is worth noting that John Smeaton, one of the founders of an independent engineering profession in Britain, owned a number of books, either edited or written by architects in the Palladian tradition; Vitruvius in Claude Perrault's edition of 1684, *The Architecture of L B Alberti* by Giacomo Leoni (1755) and William Chambers' *Treatise on Civil Architecture* (1759).[2] Their illustrations would have exposed him to a methodical and sober classicism both in their architectural style and the way in which the structure of buildings was explained through sequences of precisely engraved sections and eleva-

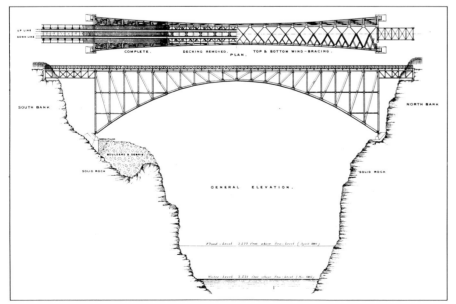

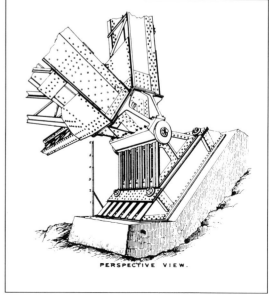

L TO R: VICTORIA FALLS BRIDGE, *c*1906, GENERAL ELEVATION AND STRUCTURAL DETAIL

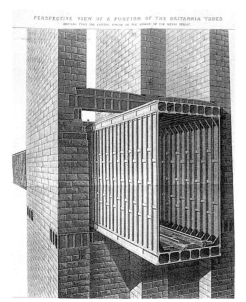

L TO R: ROBERT STEPHENSON, BRITANNIA TUBULAR BRIDGE, *c*1850; CONSTRUCTION JUNCTION AND ANGLESEY ENTRANCE

tions. That an engineer found such sources useful is hardly surprising. Architecture and what later came to be known as civil engineering were, until the late 18th century, essentially the same discipline. Together they formed part of what were regarded as the 'designing arts' – disciplines characterised by their use of drawings, models and mathematical calculation as means for constructive forethought.

What I shall call the Palladian tradition of drawing emphasised pictorial means, especially the use of case shadows, to give a greater illusion of reality. It was particularly suited to explanatory drawings and to a derivative, the presentation drawing; a work intended to impress clients and patrons, and record a project in its finished state. In this form it continued to be used by engineers, with the addition of colour and an increasing number of standardised symbols, until the end of the 19th century, and in more traditional industries until the beginning of World War I.

Drawing the Machine
Architectural drawing provided engineers with a substantial part of their graphic language but other sources were also important. The depiction of machines is an area where the conventions of architectural drawing could provide only partial solutions. From the late 15th century, attempts to describe the interrelation of moving parts led artists to develop a variety of graphic innovations. Leonardo's sketchbooks contain numerous descriptions of both practical and fanciful machines drawn with an extraordinary grasp of explanatory technique. Among his drawings which seek to elucidate the elements of mechanics is one which uses an exploded view to describe each part and then an assembled view to show the complete mechanism. But these drawings made little impact in Leonardo's lifetime and their dispersal ensured that they remained largely unknown until the 19th century.

When the artist Rudolf Manuel Deutsch used an exploded view in Georgius Agricola's treatise on mining and the metal industries, *De re metallica* (1556), it can probably be assumed that the idea had evolved as part of the general response of artists and engineers to the desire for more accurate representation. Thus it is unlikely to have been solely Leonardo's invention. Agricola's treatise is also notable for some of the earliest uses of the cutaway to reveal areas which in reality would be

L TO R: ROBERT STEPHENSON, BRITANNIA BRIDGE, SECTION; HECTOR HOREAU, SUBMARINE RAILWAY BETWEEN ENGLAND AND FRANCE, *c*1851

FRANCIS PUGH

31

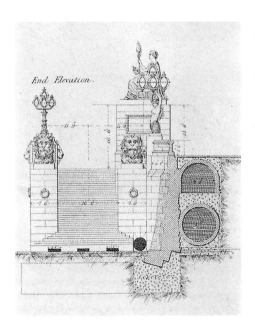

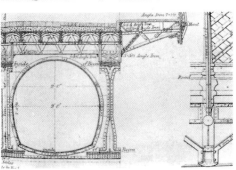
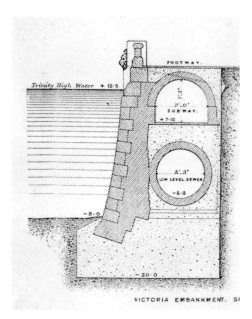

L TO R: VICTORIA EMBANKMENT. SECTION: OUTFALL BRICK BARRELS. SECTION. *BELOW CENTRE:* CROSSNESS PUMPING STATION, SEWER SECTION

hidden and the numeric or alphabetic key linking descriptive text to the parts of an illustration. Both were devices which soon became part of the standard repertoire of the illustrator of technical treatises and from this source were eventually incorporated in everyday engineering practice.

From the 16th century onwards the growing sophistication of graphic technique can be traced in a succession of works on the mechanical sciences, especially those describing machines driven by wind and water power. The earliest drawings incorporate all the information in one picture while later examples often present different aspects of a machine or mechanical process in two or three separate pictures. The illustrators of Diderot and d'Alembert's *Dictionnaire des Sciences*, one of the most accomplished of the great 18th-century works on arts and manufactures,[3] frequently adopt a sequence of views including an orthographic projection, and horizontal and longitudinal sections; followed by details of parts and illustrations of workers engaged in manufacture. Smeaton however reverts to the earlier type of single view in his illustration of the process of constructing the Eddystone lighthouse, so that the various methods for landing and hoisting stones during the six stages of construction

can be readily comprehended in one illustration. Books of this type also reproduced the style of working drawing employed by specific crafts or trades.

Despite their increasingly sophisticated presentation, illustrative techniques had many drawbacks when it came to depicting machinery. To some engineers it was now apparent that a more mathematically precise and consistent drawing system was needed to cope with the particular problem of describing the complex geometry of machine forms. The theoretical foundation was established by the French military engineer Gaspard Monge whose *Descriptive Geometry* published in 1795 uses a theoretical structure derived from the mathematics of Descartes, Pascal, and others to bring clarity and logic to a variety of *ad hoc* techniques customarily used by stonemasons and carpenters. The technique was first applied to the design of fortifications and was consequently regarded as a state secret for the decade or so prior to its publication. However, Monge eventually persuaded the French government that his system could provide much needed stimulus to manufactures and this led to its incorporation in the curriculum of the Ecole Polytechnique, an institution which Monge himself had founded. After

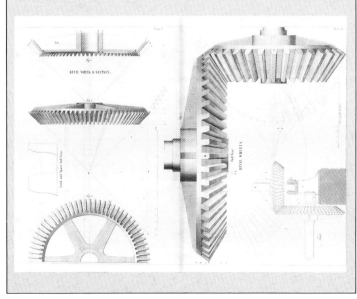
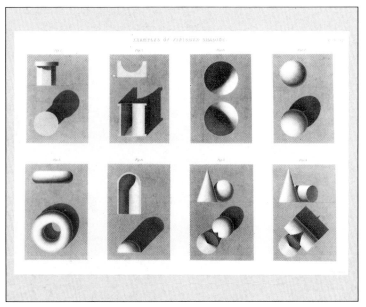

L TO R: METHOD OF DRAWING BEVIL WHEELS: EXAMPLES OF FINISHED SHADING. (BOTH LE BLANC AND ARMENGAULD *THE ENGINEERS AND MACHINISTS,* 1847)

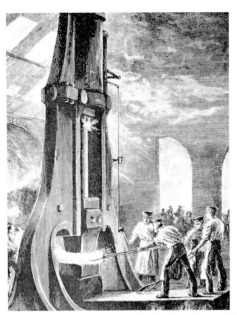

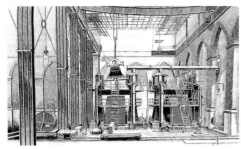
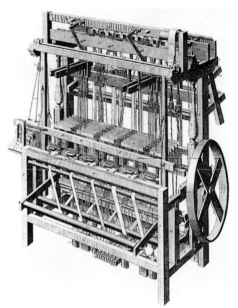

L TO R: NASMYTH, STEAM HAMMER, *c*1851: FERRANTI: DEPTFORD POWER STATION, *c*1889; RIBBON-MAKING MACHINE (FROM *DICTIONNAIRE DES SCIENCES*)

1815 it was taught in the new technical high schools established in the German states, while *émigré* French instructors at the West Point military academy introduced the theory to America. More general acceptance, in Britain especially, came only with the publication of the first textbooks on machine drawing in the 1840s. Many of these were written by French authors. The sole aspect of Monge's theory to be widely adopted in his lifetime was the standardised relationship of plan, side and front views in the form known as 'first angle projection' – a system which rapidly replaced older forms of oblique projection and other casually related views. It soon became the standard method for organising general arrangements, the type of working drawing that serves as the master to which all others relate and thus a key constituent of the engineer's design vocabulary.

Ships, Maps, Surveying and Instrument Making

This investigation would be incomplete if it concentrated on theoretical developments to the detriment of those aspects of drawing, like the majority of British examples, which evolved more or less empirically from the practice of a variety of existing trades. Those who were most influential in this respect, other than architects and builders, were shipwrights, surveyors, map makers and instrument makers.

The basic form of ships' draught had first appeared in the late 16th century in response to the shipwrights' need to predict the shape of new types of ocean-going, gun-carrying vessel. By the mid 18th century the evolution of ships' draughts, and of the mathematical calculations associated with them, had arrived at a description of the irregular solid of a ship's hull which dispensed with oblique projection in favour of a series of vertical and horizontal sections superimposed in line only on the three principal views. In British naval draughts this resulted in partly schematic, partly pictorial representation, which left much to the discretion of the shipyard foreman. A typical lines plan employs a simple colour code; black for the outline, red for the decks and fittings that formed part of the internal structure of the ship, and green for those lines like the hull contours which describe form rather than physical structure. Only works on naval architecture like Hendrik Chapman's *Architectural Navalis Mercatoria* (1768) which were intended principally for wealthy subscribers, would include projected drawings in addition to a simple lines plan; the process being too laborious and time-

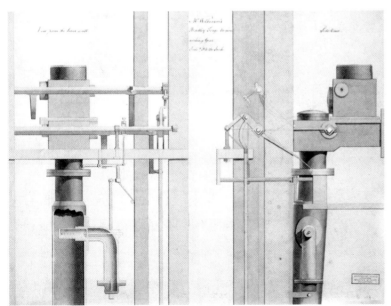
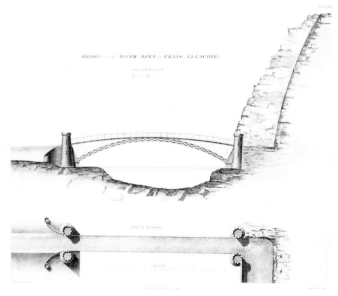

L TO R: WILKINSON'S FORGE ENGINES (DRAWINGS, JAMES WALL AND JOHN SOUTHERN, *c*1782); CRAIGELLACHIE BRIDGE OVER THE RIVER SPEY

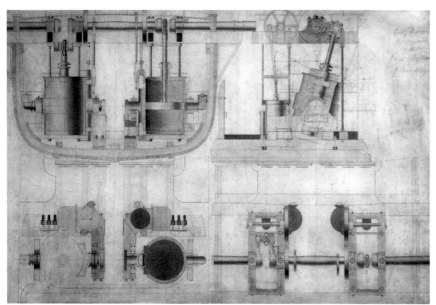

L TO R: I K BRUNEL, DRAWING FOR PROPOSED SS *GREAT BRITAIN*; LEONARDO DA VINCI, TRANSFORMATION OF ALTERNATING TO CONTINUOUS MOTION

consuming to be part of everyday shipyard practice.

The technique of land surveying by triangulation began to develop in the 15th century but accurate map-making was not possible until the invention of the first surveying instruments with telescopic sights in the early 17th century. Even then instruments were hardly precise and it is not until the mid 18th century that accurate surveying became a reality. It is no accident that the makers of the finest surveying instruments were also innovators in the field of drawing. The trade card of the London instrument maker George Adams lists instruments for copying drawings, and camera obscuras for making perspectives in addition to the more standard surveyors' equipment of theodolites, water levels, and measuring wheels. Map-making in consequence began to abandon earlier pictorial styles for a more precise plan view often with fine hatching to indicate hill shapes and a range of conventional symbols. Engineers needed a thorough knowledge of map-making and surveying, thus a familiarity with the equipment and methods seems to have played an important part in the careers of many engineers. Smeaton and Mark Brunel were trained as instrument makers, and both retained habits of precision and accuracy derived from

their early experience. Drawing, surveying, map-making and engineering are linked even more closely in the career of the topographical artist Paul Sandby, who trained as a draughtsman in the Board of Ordnance Drawing Room at the Tower of London before taking part in the Military Survey of Scotland between 1746 and 1751, and was later engaged one day a week as drawing master at the Royal Military Academy, Woolwich – the first such government appointment in Britain. The influence of his methods and style of drawing are apparent in artillery manuals, reports and other military documents of the period.

Engineering Drawing

It would be foolish to suggest that drawing and theoretical knowledge at any stage replaced the craftsman's skill at working with materials. Throughout the 19th century much emphasis was placed on the need for all engineers to acquire practical experience in addition to learning how to draw. But the evidence of collections like the Boulton and Watt archive suggests that substantial changes in working methods took place in the crucial decades between 1760 and 1790, particularly in industries associated with the steam engine and iron replaced wood as

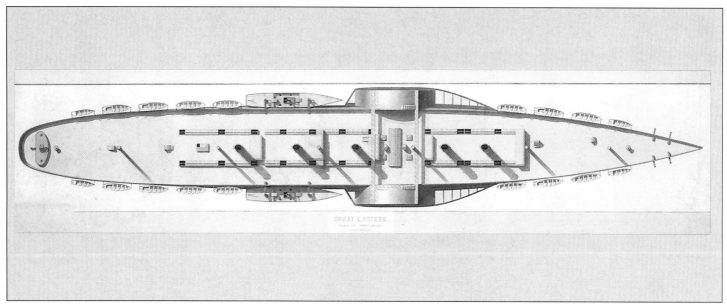

I K BRUNEL, ELEVATION OF SS *GREAT EASTERN*, *c1857.* (DRAWING BY SCOTT RUSSELL)

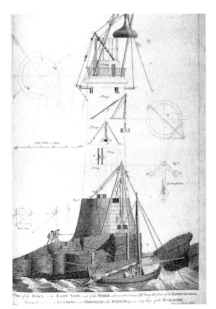

L TO R: JOHN SMEATON , EDDYSTONE LIGHTHOUSE, 1856; ONE OF THE FIRST TECHNOLOGICAL BOOKS, AGRICOLA'S *DE RE METALLICA*, c1556

the principal construction material. The pace of technological change was leading to a new form of industrial organisation in which older craft disciplines were gradually subordinated to a controlling hierarchy of engineers and entrepreneurs. James Watt and his assistant John Southern were in the forefront of these developments when they produced drawings to control and record each aspect of the engine building activities of the Boulton and Watt manufactory; a situation which in scope and ambition anticipates the central importance attached to the drawing office in British industry after 1830.

The growing specialisation that brought an independent engineering profession into being was also responsible for giving drawing its new industrial prominence; and in turn the chief draughtsman, a figure who assumes growing importance as the 19th century progresses, became indistinguishable from the industrial designer of today. It was to facilitate these changes that drawing itself began to change. Borrowing extensively from each of the sources I have described it gradually became more standardised until by the close of the 19th century it had largely abandoned pictorialism for more schematic forms of representation which were less ambiguous and capable of more precise interpretation. From 1870 onwards the descriptive function of drawing was challenged with increasing success by photography. In contrast, its importance as a tool for prediction and analysis continued to grow. Perhaps one of the more thought-provoking aspects of the computer drawing systems is the possibility they offer for reintegrating pictorial depiction back into the design language. The gain in perceptual clarity would be especially beneficial to clients and users as well as to designers themselves. Then, a fuller understanding of the origins and developments of the language of drawing becomes more than a matter of passing interest. A closer study of the experience of earlier generations of artists, architects, engineers and designers is essential in attempting to define new forms of visual language appropriate to our immediate needs.

Notes

1 James S Ackermann, *Palladio*, 1966, pp 171-172

2 A W Skempton, (ed), *John Smeaton, FRS*, Appendix I, 'Smeaton's Library and Instruments', 1981

3 The *Dictionnaire des Sciences* was published by Diderot and d'Alembert for the Academie des Sciences at Paris.

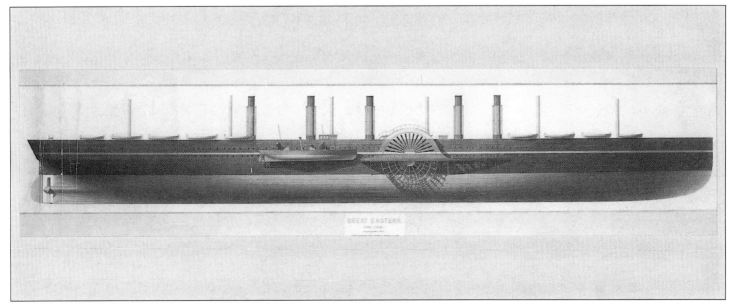

L TO R: I K BRUNEL, PLAN OF SS *GREAT BRITAIN* (DRAWING BY SCOTT RUSSELL)

Photo: Harry Sowden

RICHARD ROGERS & PARTNERS, LLOYD'S OF LONDON, DETAIL

THE AESTHETICS OF ENGINEERING
Charles Jencks Interviews Jack Zunz

Photo: Otto Baitz

RICHARD ROGERS & PARTNERS. PATSCENTER. PRINCETON

The consulting engineering practice Ove Arup & Partners prides itself on a creative involvement with architecture. In a major interview Jack Zunz of Ove Arup and Charles Jencks discuss Arup's contribution to contemporary architecture from the Smithsons' Hunstanton School to Foster's Hongkong and Shanghai Bank and explore the aesthetics of engineering in the aftermath of an era when engineering itself was celebrated as the paradigm for an architectural aesthetic.

The Decentralised Multinational or Controlled Anarchy

CJ I'd like an overview of Arup's. You have, I understand, more than 3,000 staff in 28 countries, which makes you something of a multinational. How do you try to control that?

JZ We do not control it on a day-to-day basis in the strict sense of the word. It all started with Ove Arup – a very unique person whom I won't even attempt to describe to you here. What is relevant is that Ove's style was to let people get on with their own thing, provided he trusted them. And that is something which has caught on and continues to permeate our whole firm. We are of course, a firm which is heavily design-orientated and we tend to let our young people, who are generally carefully chosen and more often than not home-grown, develop themselves to their full potential doing their own thing, all within broad guidelines and philosophies. For instance Australians have come to us here in Britain and have then gone back home but stayed with us. Now our Australian offices are independently run by Australians, many of whom have been with the firm for more than 20 years.

So, yes, we are a multinational, but we're essentially a number of groups of designers, based in different towns and countries, who all share common ideals and objectives. You see, the way we associate with one another is also unique. Nobody in Arup's owns any part of the firm. We don't have shares. We have given the equity of the firm to employees and charitable trusts. In having a large group of people practising together professionally we have attempted to remove financial tensions so that everybody in Arup's knows that all the money which comes into the firm *stays* there. And we share it out amongst ourselves, though some get a larger share than others, obviously. We invest in education, training, research and development, and we sponsor young students.

CJ That sounds unique in an engineering company. One question I want to ask: who could you be compared to as an engineering firm? What companies have you modelled yourself on?

JZ We don't model ourselves on anyone. We've worked out our own way of doing things. I suppose there must be other firms which have similar objectives and it would be interesting to compare ourselves with them.

CJ I don't know the background, but is the set-up of the firm Scandanavian?

JZ No, not at all. It started off as a traditional partnership. Arup was actually born here in Newcastle-upon-Tyne, the son of the Danish Consul, but was educated in Germany and Scandinavia. He first worked in Hamburg for Christiani & Nielsen, the Danish firm of Civil Engineers, Designers and Contractors, before returning to this country in 1923. In 1946 he founded the consulting engineering practice, and three years later took Ronald Jenkins, Geoffrey Wood and Andrew Young into partnership. This continued into the 1950s when Peter Dunican was taken into the partnership. And then gradually, as the firm grew in size and complexity, certain ideas or philosophies began to evolve. By the time Povl Ahm and I were taken into partnership in 1965, a number of things had become clear. Two of the more important ones were: first, even though we were partners, we

FOSTER ASSOCIATES, STANSTED AIRPORT PASSENGER TERMINAL ROOF

didn't own any equity in the group – we paid nothing to get in and get nothing when we go out; the second principle was that we were all equal.

CJ What would you call this if you gave it a generic term? 'Corporate' what?

JZ There is no label. Ultimately the Ove Arup Partnership is a singular company – a partnership for the benefit of its members.

CJ Is there another firm of engineers which is your competitor? If you don't get the job, who does?

JZ Of course there are a lot of competitors. The firm is broken down into a series of more or less self-governing groups or villages covering a very wide spectrum of expertise. Our competition tends to be . . .

CJ Yourself?

JZ Our groups do tend to compete with one another. For example, in a number of major architectural competitions we've found Arup Associates up against Norman Foster, Richard Rogers, Jim Stirling and other leading architects supported by one or other of our engineering groups.

CJ You can't lose?

JZ Well, you do sometimes. In fact, it's very healthy to do so.

And fortunately, I think people have come to accept that the group helping one architect will be so keen to win that they will not in any way compare notes with others in the company.

CJ It's very strange, almost a conflict of interest situation. And you are a monopoly, but you have such a wide service that you can accomplish things that other people can't, right?

JZ We do have in our midst international experts, all sorts of specialists, who are not available to others, that's true. But to answer your question: while we do sometimes compete with ourselves in certain fields, we mainly compete with others. We admire some of them very much indeed, and in fact a number of them originally came from Arup's.

CJ Who do you admire, for instance?

JZ I suppose in the recent past those I admire very much are people like Peter Brett, Peter Campbell, Tony Hunt – but then of course there are great firms like Freeman Fox, Maunsells, Scott Wilson and many othes.

Brutalism and Modernism

CJ If you look at the book celebrating your 40th anniversary you find it reads almost as a roll-call of famous British architects.

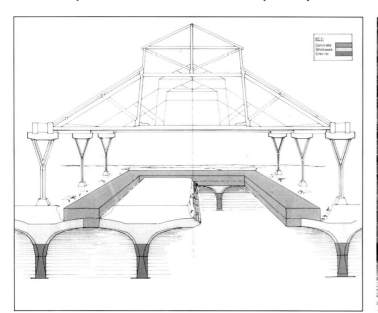

TERRY FARRELL PARTNERSHIP, TOBACCO DOCK

A & P SMITHSON. HUNSTANTON SCHOOL

I'd like to go through some of them and ask your opinion of such things as the Smithsons' Hunstanton School.

JZ I did a little work on that myself.

CJ It's obviously in the Modernist tradition, and also beautiful in some respects.

JZ So the architectural press have reported.

CJ But one wonders if it's at all good for children. Does it look like a school, function like a school, and is it the kind of space that children like to be in? I'd like your opinion.

JZ I can't really answer that because although I was involved in the school's design and construction, I was out of the country for six years after its completion. I've never visited it while it was functioning so really whether or not it works as a school is not for me to answer.

CJ What about the *idea* of the school?

JZ At the time it appealed to me very much indeed. When the plan was conceived and the school was built there was so much chaos generally that the order in the nature of the building was something that struck a sympathetic chord.

CJ But the idea of a school is not necessarily these reverberant spaces?

JZ A lot of the schools at the time weren't as well thought out intellectually as this one, although many of the schools that followed, in the 1950s, had a not dissimilar arrangement.

CJ But do you accept that it might be good architecture and bad building, if one makes such a distinction?

JZ Yes.

CJ Because, it's the first Brutalist building to be both harmonious and well ordered.

JZ You put the word 'Brutalist' next to it.

CJ That's not me, that's historic.

JZ There may be some confusion here. At the time – 1950/51 – we were designing it, the word 'Brutalism' hadn't been invented. The concept of the building was strongly influenced by Mies's work, so why is the word Brutalism attached to the Hunstanton project but not to Mies's work?

CJ Because of the tower, the use of industrial elements, the open drains in the loo. Because of the hard surfaces, the tough industrial realism. That's the essence of Brutalism, although I agree the word wasn't around then.

JZ There's another point you made. You said that we had been involved with a great deal of architecture in this country

L TO R: RENZO PIANO. MENIL GALLERY; STIRLING WILFORD & ASSOCIATES. STUTTGART STAATSGALERIE. INTERIOR

Photo: John Maltby

L TO R: PARK HILL HOUSING. SHEFFIELD; LUBETKIN & TECTON. ROSEBERY AVENUE FLATS. LONDON

since World War II. Now that's perfectly true. Ove Arup became interested in Modern architecture in the 1930s, and he was a very eminent member of the MARS Group. The whole idea of engineering as a more creative, more positive contribution to the total building process was something which Ove and his collaborators introduced into our industry. And that of course raised the whole level of engineering in the building profession. This is the reason why, I think, we became involved with so much significant post-war work.

CJ Of course that allies you with Modernism, like it or not.

JZ I'd rather say with contemporary architects and architecture.

CJ Let's go right to the heart of the problem: the Park Hill, Sheffield, housing scheme, and schemes of that ilk, which come out of Brutalism and represent, if you like, the unacceptable face of Modernism. What do you say about those schemes, and your involvement and culpability in their invention?

JZ I don't think we are culpable. If these schemes are unacceptable, and don't forget there was a great deal of theorising at the time which seemed to indicate that this was just the type of housing that was needed, then surely sociologists, clients, planners and architects must be responsible. Our contribution is engineering; the decision as to what is to be built is not really in our hands.

CJ You can't wash your hands of the effect. I mean, shouldn't you at a certain point say: this is socially unacceptable?

JZ Well, no. In fact the general consensus was quite to the contrary. At the end of the Second World War, we were all infused with a euphoria. My daughter recently said to me that she was really sick of the way television glorified war: the only thing that bothered her was that in World War II there really seemed to be something evil that people were right to take up arms and go to war against. And of course we felt that very deeply at the time. When we started practising – I started in 1948 – we were still excited by the idea of helping to create a wonderful new world, and technology was there to be harnessed in the service of humanity. It was an idealism, in that sense, embodying some of the ideas which the Modernists, as you call them . . .

CJ Well, as you call them too. It's not just me.

JZ I didn't use that word. I think everything contemporary is modern.

L TO R: RICHARD MACCORMAC. FITZWILLIAM COLLEGE: PHIPPEN RANDALL & PARKES. MIDDLESTONE. ST BRIDES. LONDON

L. TO R: SPENCE. BONNINGTON & COLLINS. HAMPSTEAD CIVIC CENTRE: LESLIE MARTIN & COLIN ST JOHN WILSON. LAW LIBRARY. OXFORD UNIVERSITY

CJ   Hold on a minute. You've admitted that there is an ideology involved in this – an ideology of Modernism.

JZ   It was *idealism*: the ideal of creating a just society. We really thought that the vision of people like Le Corbusier was going to be put into practice for the benefit of all. We tackled these problems with great enthusiasm because we thought that by contributing to these buildings the community as a whole would benefit and all would live happily ever after. And most architects were totally committed.

CJ   In good faith maybe for five years. In fair faith for the next five years. And then, as the 1960s progressed, the ideals became ideology. In any case your firm and probably you yourself have done a lot of Modernist work with Basil Spence, Leslie Martin, ABK, and now Rogers and Foster. And in so far as you provide a backdrop for these other people (or maybe even a crutch or a creative helping hand or whatever you want to call it) you have really been the Modernists behind the throne, as it were. Would you really deny that?

JZ   Yes, absolutely, what we've done, as engineers, is help people create whatever they wanted to create. You haven't mentioned Tecton. We started with Tecton really – and then

Maxwell Fry. But we also worked with T P Bennett, and he can hardly be called a Modernist.

CJ   Yes, but what I'm trying to get you to admit is that there's a kind of ideology or world-view of Modernism in which you are a leading force. It seems to me that it's silly not to admit this, and see both its advantages and disadvantages.

JZ   I think that while the Modernism you're talking about was part of the Brave New World which we all envisaged after the war, we found that, as architecture matured in the 1960s and 70s, whatever idealism we had was funnelled into good building, good architecture.

CJ   But whether we put an 'ism' to it or not, whether we admit it or not, we see with hindsight that certain distinct choices were made, though I agree that it can be looked at simply in terms of good building and bad building.

JZ   Good designers don't sit down with a blank sheet of paper and say self-consciously: 'Well, I'll do this in such and such a style.' I think that what they design is more a response to inner convictions, which are not necessarily always articulated. And I suppose that the more we, as engineers, acquire an interest in architecture, the more we'll be able to relate to the architects

Photo: Archie Hanford

L. TO R: LUBETKIN & TECTON, HIGHPOINT, HIGHGATE: T P BENNETT & SON. SMITHFIELD MARKET

Photo: Barry Dunnage

Photo: Barry Dunnage

RICHARD ROGERS & PARTNERS. PATSCENTER. PRINCETON

with whom we work. I mean that the delight we create comes at the ultimate behest of the architect.

CJ But are you washing your hands of the delight?

JZ No, it's nice to be party to it. I don't think there's enough delight.

CJ If I understand you correctly, Arup's occupy a strange position with respect to other engineers because you're more creative: you suggest things to architects and that must create some of the delight.

JZ We hope so, yes.

CJ So can we talk for a minute about the aesthetics of engineering. This is often discussed in terms of truth leading to beauty, with simplicity and economy being the key to creating a good solution to a complex problem. Do you talk about that explicitly at Arup's, or is it something you pick up by osmosis?

JZ Well, we talk about it from time to time, but I think that underlying our work is a freedom or non-doctrinaire attitude to the problem-solving process. Take Hunstanton, which we touched on earlier. It is simplicity incarnate, and it has, to me, a kind of beauty. However, you can also achieve beauty with very ornate designs. And sometimes the logical and the simple can be

very ugly. That is why ultimately all architects need to have some artistic guidance.

CJ I'm delighted that I've heard these words coming from you because they are very rare words from an engineer.

JZ Yes. Many people don't agree with them.

CJ Do you hold lectures on aesthetics in the office? Do you study the history of architecture?

JZ Individuals do. And then a while ago we initiated a series of lectures which continued over a period of two years. All the current greats – Jim Stirling, Norman Foster, Richard Rogers – came and described their work. We've gone out of our way to try and expose our people to things other than computers and the more hard-edged tools of our trade. We want to expose them as much as possible to real artists as well. This notion is quite foreign to formal engineering training, and indeed to the chartered, learned institutions to which we all aim to belong.

The Late-Modern Aesthetic

CJ I'd like to look at one aesthetic which is apparent in a lot of British work, the use of tensile structures: things such as Rogers' Fleetguard and Patscenter, Hopkins' Schlumberger, or Foster's

Photo: Harry Sowden

MICHAEL HOPKINS ARCHITECTS. SCHLUMBERGER FABRIC ROOF. CAMBRIDGE

Photo: Harry Sowden

Photo: Harry Sowden

FOSTER ASSOCIATES. RENAULT CENTRE. SWINDON

Renault Centre. In all of those buildings there's an exaggeration of the taut structure – a kind of exo-skeleton, carried out above the building and painted in primary colours. The Patscenter, for example, had a kind of Pop art ring which it features in cables; Fleetguard has red struts in a kind of cat's cradle mesh; Schlumberger looks like a spiderweb, with black struts holding up a rather punk-style tent, or skin, pulled out in an excruciatingly painful manner; and Renault looks like a lot of umbrellas in an awful Renault Yellow.

What strikes me about the aesthetics of this is that there's no real theory of composition in a traditional sense. They start from a modular idea and diminishing or augmenting them is a matter of the sensibility of the architect. Now, in a traditional, slower-changing society, you would find a discussable theory of composition coming out of a school of very expressive, creative architects and engineers working together. You'd formalise the *rules* of the game. What would you say to that?

JZ I'm not sure whether I understand you. Certainly, some of the materials we work with now – particularly steel and especially some of the tensile steels, but even concrete and pre-stressed concrete – give us, and the architect, a great deal of freedom.

The 19th-century designer had a much more limited vocabulary to work with. And the consequence of that is that there's bound to be a lack of order, as you put it.

CJ I'm not saying that there's a lack of order, but that there's a lack of a theory of composition.

JZ I think these buildings nonetheless all contain a logic, a rationale. Take, for instance, the great rings which are part of the aesthetic of the Patscenter. It is technically possible to make that junction in another way, but the ring is quite an elegant solution because it's a very strong element.

CJ But it represents a kind of hand-crafted, Anthony Caro-like beauty of technology. Now I'm going to argue for a second that placing such a thing over the building and painting it in a brash 1960s colour is a very un-Modernist thing to do in the sense that Gropius, Corb, Mies always maintained that a building had to be balanced. They would have seen this as structural exhibitionism. How would you reply to that?

JZ I've given up trying to comment on how architects describe each other. In nearly 40 years of practice I've rarely heard one architect say nice things about another.

CJ But aside from name-calling, the Modernists tried to put

Photo: Richard Bryant

L TO R: MICHAEL HOPKINS ARCHITECTS. MOUND STAND. LORDS; FOSTER ASSOCIATES. WALDSTADION. FRANKFURT

EDWARD MILLS, HALL 7, NATIONAL EXHIBITION CENTRE, BIRMINGHAM

forward the idea of a generalised, well-behaved 'taste-culture', with standardised, harmonious, products.

JZ I think Richard Rogers is probably a romantic, because he uses the tools of technology in a very decorative way. His style, whether you like it or not, is a very personal one.

CJ I personally like it, but with a certain reluctance, because in it the structure takes over, and I can think of many other things for the building to express.

JZ The fact is that the structure is actually used as a means of architectural expression, not only in its overall form but also by careful manipulation of all details.

CJ But don't you think it's wrong when the exoskeleton and pipes are placed in the foreground, like prostheses? And isn't exposing and hand-crafting all of that High-Tech very expensive at the end of the day?

JZ Old Gothic cathedrals have buttresses externally.

CJ But the designers of a Gothic cathedral didn't set out to express structural elements: the buttresses came about as a consequence of creating those very high spaces. And it seems the other way around here: the structure seems to be the reason the building was created. Now, knowing how much trouble this involves, we also know that the architect and the engineers conspire together to do this.

JZ Have you heard of Edward Mills' Hall Seven, at the National Exhibition Centre? That has a suspended structure, for absolutely logical reasons – though it may not be a rationale which you enjoy. You can't expect us as engineers always to be in complete sympathy with the architect. We've worked on a number of schemes where the architecture has been awful – for me at any rate.

CJ I'm trying to make you respond to the difference between the Hunstanton School with its Miesian moralism, and this Late-Modernism.

JZ Well, I won't in the way you want me to. Starting from Hunstanton to where we are now, there has been a steady technical progression where different architects use available tools in different ways – whatever 'ism' is attached to the result is, I'm sure, interesting to many but bewildering to me.

CJ So you wouldn't accept that it's 'Late-Modern' to exaggerate structure in this way?

JZ It may well be so judged by historians in years to come, but I don't see it like that.

L TO R: COVELL, MATTHEWS WHEATLEY, ROPEMAKER PLACE; ANDREWS INT & ANDERSON NOTTER FINEGOLD, INTELSAT; PALMER & TURNER, EXCHANGE SQUARE

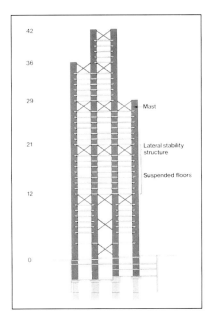
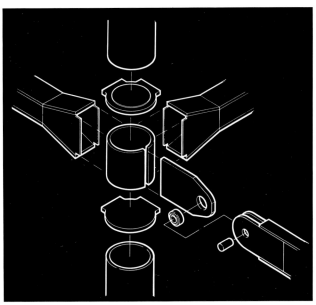
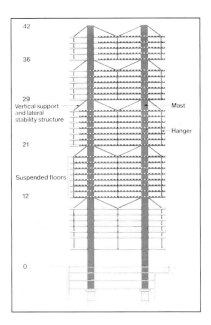

FOSTER ASSOCIATES. HONGKONG BANK

CHARLES JENCKS/JACK ZUNZ

The Hongkong Bank

CJ Can we go on now to the Hongkong Bank, where an incredible proportion of the cost of the building was taken up by the structure and the cladding – some 58.4 percent of the total US$640 million. Now, for Norman Foster, the space, openness and flexibility this created justified the greater cost, but on another level – and this concerns the ethics of engineering – it doesn't seem to be very much like Fuller's argument of doing more with less: you're doing more with more. Can you really justify the cost of the structure in terms of the building's ultimate flexibility, or are you not, like the builders of the pyramids, simply doing it because it can be done, and because you wanted to try?

JZ The answer's obviously no. We, as engineers, wouldn't just do it because it's there to be done. And you ask whether such an expensive building is justifiable. Ultimately, that's for the client to say, although the total cost of the Bank, fitted out, is not higher than other, similar, totally fitted-out buildings for different corporations and banks. The costs on the bank are always compared with those of other steel and core buildings – not comparing like with like. I also think we have an obsession with money, even when it's totally irrelevant.

CJ Well, no it isn't.

JZ But it *is* irrelevant. How can you quantify in dollars or pounds people's enjoyment – particularly when you are doing something you can well afford to do?

CJ The reason why we do it is that money is the only symbol system we can use for comparison.

JZ But if we keep on doing that, we will never create anything worthwhile in our society. The great buildings which we worship were all created without accountant's yardsticks like percentage return on capital, etc.

CJ There were other kinds of accountants. Mohammed said that architects were the ruination of a kingdom. Throughout history, people have always complained. You have to make those comparisons all the time and say, this amount of money will buy you that kind of architecture. Some of the money spent on the structure of the Hongkong Bank could have been spent on the fabric, or on paintings . . . like Van Gogh's *Sunflowers*.

Now, the Hongkong Bank is a corporate headquarters, and you could say that the corporate headquarters, along with the museum, has become *the* 20th-century building-type. In a sense,

FOSTER ASSOCIATES. HONGKONG BANK

45

Photo: Max Dupain

Photo: Max Dupain

UTZON/HALL TODD & LITTLEMORE, SYDNEY OPERA HOUSE

you and Foster are directly involved here in an ideological expression of what is held as most important in our culture. There's a moral responsibility to admit what is, in fact, the case – that you are spending a lot of money on it and that it's a symbol of our time. Then comes the inevitable critique: many people will find it banale to spend so much money on banks and the expression of structure, and given the opportunity, politically, they would spend it otherwise. . .

JZ I agree that's the case, but I'm not sure that I go along with the jury test: guilty or not guilty – what's good or what's bad. Who is to decide? Would you give the Sydney Opera House a verdict of good or bad? And how do you vote on the Centre Pompidou?

CJ Well, both are cultural centres, and as such, it seems to me, have more general validity than a bank. But you could say that for the people of Hong Kong, it is a bank which has become a symbolic focus . . .

JZ I think it's too early to say; we have to wait a few years. But I suppose the question really is: are we justified in spending this money on the structural expression which gives this building its particular architectural pleasure?

CJ Well, it escalates the muscle-power problem. If you're one of the 4,000 Chinese or 200 foreigners who work in the Bank, you feel pride in it. But think of all those other competing banks and institutions like the China Bank, who then have to try to outdo it.

JZ You're implying that I can do something about this.

CJ But everybody can't wash their hands. I'm going to ask you a question directly: is there a building which you would refuse to do?

Ethical Questioning

JZ We've often debated when we've been working on buildings with architects whether we believe we're being led into the wilderness, and it's a very difficult problem.

CJ Well, can you give us some guidelines, and indicate what your answers might be?

JZ The Opera House is a good example, because we thought at one stage that Utzon was probably leading us astray, as he wasn't coming up with the goods. We had a client who was very indulgent. But we did support him right to the end, although there were times when we wondered whether were doing the

RENZO PIANO & RICHARD ROGERS, POMPIDOU CENTRE

L. TO R: PHILIP COX & PNTRS, YULARA VILLAGE, AUSTRALIA; DR K EL-KALAWI, DOHA UNIVERSITY, QATAR

right thing.

CJ In architecture every school is attacking another kind of school, buildings are being stopped. Do engineers have these ethical battles, or battles over style?

JZ Yes, we have a lot of ethical battles. For example a number of my colleagues won't work on any military establishments. Within the firm, as I said earlier on, we have a form of controlled anarchy, and we are sensitive to our people's ethical tolerances.

CJ But apart from the building types and the regimes for which they're built, there are ethical issues directly related to engineering. An engineer can make anything stand up more or less, given enough money. But is there a point at which you say: well, this is an upside-down pyramid and I refuse to do it?

JZ Well, yes, we refuse, and normally, sensible architects will then try to find another solution. Certainly in the crazy 50s architects used to come to us with schemes which were totally unbuildable, and we said: 'We won't do it. It can't be done!' so they'd start again.

CJ Can you give me an example of a building which you've refused, on the grounds that it was too uneconomic, or bad engineering. Because in this discussion, we're trying to draw

guidelines and, you know, I can't see them.

JZ Probably because there aren't any guidelines in the way you want them. What I'm trying to tell you is that I don't believe you can legislate or draw up a code of conduct for integrity in engineering at any rate. I think integrity is something which you either have or haven't got. If you try to codify it, you soon end up in trouble. You ask me to give you a particular example where we've walked off a job. I suppose I could but in any case I don't think there are any which would necessarily interest you, because they haven't had a sufficiently high profile.

CJ I'm trying to draw you out so that someone outside your field can see the kind of things you grapple with in your distinction between good and bad engineering, as applied to architecture, because clearly you must have that distinction.

JZ Well, this is where I'm not sure if you and I are really on the same wave-length. When we've talked about Patscenter, Renault or the Hongkong Bank, we've talked about the structure. I believe (and I may be biased) that the engineering in these buildings is very good. You want me to distinguish between good and bad engineering as applied to architecture. It is possible to discuss structure in its context – is it appropriate, is it

OVE ARUP & PARTNERS. KYLESKU BRIDGE (*LEFT*); CHEPSTOW BRIDGE (*RIGHT*)

elegant, well detailed and so on. If it gets good marks on most of these counts – and some of them are of course very subjective – then I suppose that could be described as good engineering. Whether the architecture is any good is another matter. Fortunately, engineers haven't got round to describing their work in 'isms'.

So, when we talk about the building as a whole, I'm quite happy to talk about the structure and how it works and so on, but ultimately its rightness or wrongness is part of the whole and this is, in a way, part of our firm's philosophy. And while, clearly, you could take a building apart and criticise it in technical terms, those criticisms wouldn't be of great interest to you, because they'd be about details and trivia in your eyes, so I can only really talk about structure in the context of the building.

CJ There's a trend towards evanescence in structure, I mean, doing more with less. Do you think that you, the engineer, might soon disappear visually?

JZ The post-war trend of building thinner walls with less reinforcement is now reversing. I think the whole issue of durability has come to the fore. We're building much more soundly now. But in a way, what we lack – and this is probably something which would appeal to you – is a basic code, in so far as we haven't a clear idea of what society wants. Do they want these things to stand for five years, 10 years, 20 years?

CJ I would have loved to have asked you about Roland Mainstone's central idea that engineers learn from their failures; and what wonderful failures you've had that you've learned from.

JZ Quite a few – probably not as spectacular as some other people's – but we've had some. But society has changed totally in its attitude towards failures. I read a marvellous little monograph on one of the directors of the Great Western Railway. This chap kept a diary during the period when Brunel was going up and down the West Country creating the most wonderful things, but also having the most spectacular failures. Every time the news came that something had gone wrong, this chap entered it in his diary and noted that the stock of the railway company had been marked down from £110 to £109, or whatever. Today people who commission buildings want certainty – the whole question of risk has disappeared and, as a consequence, there is a premium on innovation, but that's a long story.

'The Engineer's Aesthetic'

CJ A quote from Corb's *Towards a New Architecture*: 'The engineer, inspired by the law of economy and governed by mathematical calculations, puts us in accord with the universal law. He achieves harmony.'

JZ Like all these statements, it sounds great and is half right. There is some harmony in certain mathematical shapes and so on, but I think, as I said earlier, I find all generalisations and doctrines very suspect.

CJ The reason I read you this Corb quote was that it is *the* Modernist belief: the engineer as noble savage, right? And I wondered how you felt about being a 'noble savage' for architects, because that's the role the Modernists cast you in. Would you say that architects burden you with the onus of solving their problems with structure in a harmonious way?

JZ The noble savage concept is intriguing – I will now see my colleagues in a new light. Some architects come to us with a blank sheet of paper and say: 'Well, you're Arup's, what do you think we should do?' Others come with a virtually finished concept. Their minds are made up – this is what they want to do, and all they want from us is some help to make sure that it stands up, that the ventilation works, and so on. And you have all stations in between.

CJ Well, can you give us some instances: for example, how does Foster compare with Stirling?

JZ Foster often starts with an idea that he hasn't fully articulated either in his mind or on paper and there's an open discussion about how it might be achieved. It is, like so much design, often an exciting voyage of discovery. Now Jim, generally, is much more introverted in his concept. He is more likely to work on his own and then articulate a concept which has considerable special and aesthetic coherence before inviting technology to play its part.

CJ Would you generalise about any of them – for instance Rogers against the other two? How much of an input do you have in his work?

JZ We probably participate more with Rogers and Foster in the early stages than with Stirling – although there comes a point in all these schemes where our involvement is total.

CJ Do you feel that you should get more credit for these things?

JZ Some of my colleagues do – I don't really care one way or the other. I think that the credit you get is less important than what you feel about your own achievements – difficult to be honest.

CJ I think that's very altruistic. When people put a lot of their lives into a building, they surely want to get credit for it?

JZ Well, suppose most people like to be credited for what they do (and what they don't do sometimes). But I have a suspicion that people in what you might call the creative industry of architecture become corrupted by too much hype and exposure. Perhaps I shouldn't say that to a distinguished critic like you, but I think there's something in it. And I think, ultimately, no matter how often you're photographed or published, you have to know inside yourself what you've achieved and what you haven't achieved – what your strengths and weaknesses are. And so whether you are credited or not assumes a lower order of priority. Some of my colleagues are very sensible about it.

CJ In Arup's?

JZ Yes, in Arup's, but also in the profession in general.

CJ Because architecture is a prima donna activity?

JZ Oh yes, of course, it's a much more egocentric activity than engineering.

CJ But on the other hand, architects are more exposed to a lot of these pressures of self-expression. The client wants the architect's stamp on his building and our society celebrates the name brand. It's not just an accident that the profession is how it is.

JZ That is probably arguable but in any case our culture is a very anti-technology culture at the moment. An engineer is the chap you call in to fix your central heating, unlike other European countries and Japan, where the profession is held in the highest esteem. You're quite right that engineers don't have the same exposure. Indeed, when the Queen opened the Forth road suspension bridge many years ago there was an article in *The Times* that mentioned the architect – who played a very minor role, probably helping with the details – and of course the Secretary of State of Scotland, but the consulting engineers weren't mentioned at all.

CJ Well, it sounds like you're standing up for your profession now!

Arup's were the engineers on all the projects illustrated in this interview. Many of the photographs are reproduced from a monograph celebrating the 40th anniversary of the Ove Arup practice: Ove Arup & Partners, *published by Academy Editions.*

PROJECTS
Engineering in Contemporary Architecture

FREI OTTO, DIPLOMATIC CLUB, RIYADH

TERRY FARRELL PARTNERSHIP
Lee House and Embankment Place, London 50

FOSTER ASSOCIATES
Médiathèque, Nîmes 54

AHRENDS BURTON & KORALEK
Cummins Engine Plant, Shotts 58

FREI OTTO
Diplomatic Club, Riyadh 62

The relationship of engineering to architecture is a necessary and constantly imaginative one. Engineering is seen by some as a vital structural component, by others as an inspiration acting as a paradigm for an architectural aesthetic. The varied use and role of engineering in contemporary architecture can be seen in a range of current projects by leading architects Terry Farrell, Norman Foster, Ahrends Burton and Koralek, and Frei Otto who although different see engineering not as a mechanical necessity but as part of the aesthetics of architecture.

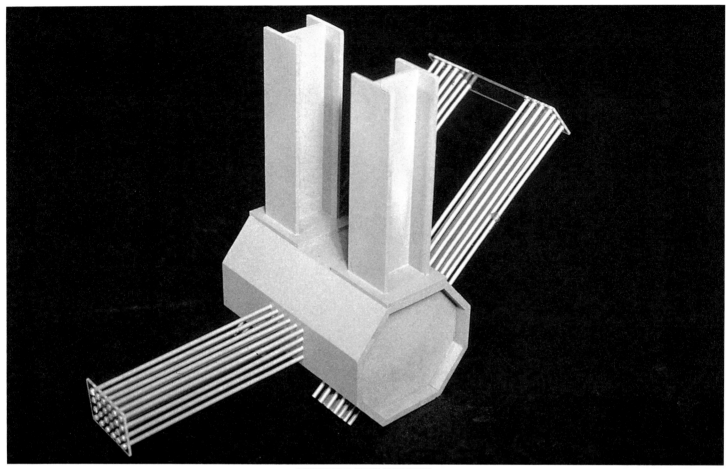

STRUCTURAL JOINT, LEE HOUSE

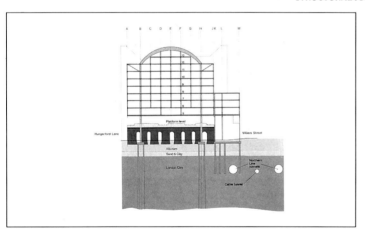

EMBANKMENT PLACE, CROSS SECTION

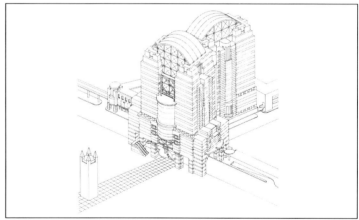

LEE HOUSE, AXONOMETRIC

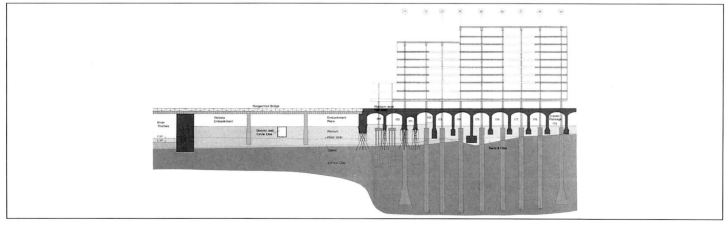

EMBANKMENT PLACE, LONGITUDINAL SECTION

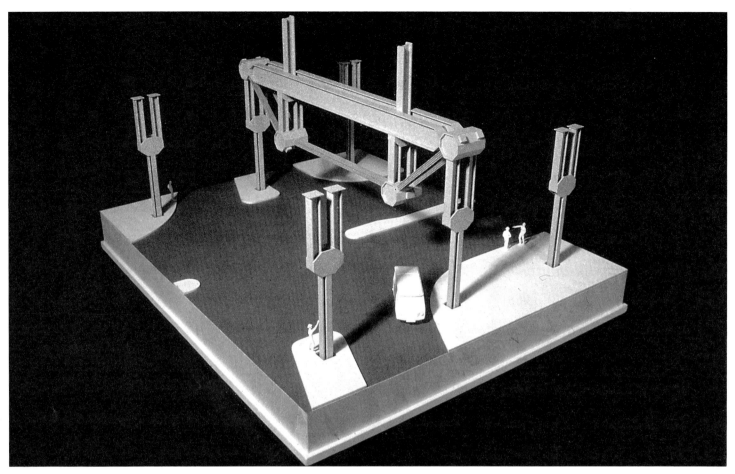

LEE HOUSE. MODEL OF STRUCTURAL SECTION

TERRY FARRELL PARTNERSHIP
Lee House & Embankment Place, London

Lee House is a two-block office complex bridging the London Wall dual carriageway. The building fulfils two primary functions: to bridge and thereby harmoniously unite the functional, leisure and pedestrian zones of this prominent urban area and to answer a desire for increased office space for a proposed workforce of 3,000.

Echoing the symbolic medieval Cripplegate that denoted the boundary of town and country, Lee House is a ceremonial gateway to the Barbican complex. A sense of symbolic stature echoing the traditions and grandeur of this historic site is mixed with a familiar concern for ornamental form, detail and colour. Lee House enhances rather than ignores the existing site: one block bridges the dual carriageway, and existing concrete walkways are to be replaced by half-enclosed bridges. Pedestrian space access and environment is respected by the creation of sympathetic, dense, variegated spaces that contrast with the original 50s schematic planning.

The office floors meet the modern practical requirements for increased space, openness and flexibility and are invested with individual character by the well-judged insertion of conservatories and service areas.

Space with identity, variety of colour and ornament, respect for urban needs and modern functional requirements as well as sensitivity to an historic site are united in the symbolic and practical functions of Lee House.

Embankment Place is a major urban scheme that seeks to regenerate the Charing Cross area by means of a sensitivity to small and large scale adjustments to a major site. The scheme consists of a new Air Rights Building to be constructed over the railway platforms of Charing Cross Station, a pedestrianisation scheme in Villiers Street, an extension of the Hungerford footbridge linking Charing Cross with the South Bank, the introduction of arcades on the west and east sides of Villiers Street, and the opening up of the existing Embankment Gardens.

The Air Rights Building comprises two building elements: the main building above the tracks and the Villiers Street block situated between the Charing Cross Hotel and Carrara House. There will be six floors of deep plan office building on the Thameside frontage extending to eight storeys towards the Charing Cross Hotel at the rear. The building is conceived in stature and size as a palace in the Thameside tradition that includes the Palace of Westminster, the National Liberal Club and Somerset House amongst others. The building's mass and form consists of an adaptation of an arched bowtruss structure. The Thameside elevation will be a formal composition with highly decorative surfaces. The two corners are intended to be solid robust elements giving a sense of solidity whilst there is a fairly light filigree structure of delicate glazing in between expressing the full contour of the bowtruss arch and reminiscent of the original station roof. A central atrium will bring daylight deep into the body of the building. The huge bowtruss is supported by columns which are in turn supported on deep piles. All the floors are hung from the trusses.

Collaborators: *Architects* Terry Farrell Partnership; *Engineers* Ove Arup & Partners; *Developers for Charing Cross* Greycoat; Photographs courtesy of the architects and engineers

EMBANKMENT PLACE. THAMESIDE ELEVATION

THAMESIDE FACADE, AIR RIGHTS BUILDING

EMBANKMENT PLACE. VILLIERS STREET ELEVATION

AERIAL VIEW OF EMBANKMENT PLACE

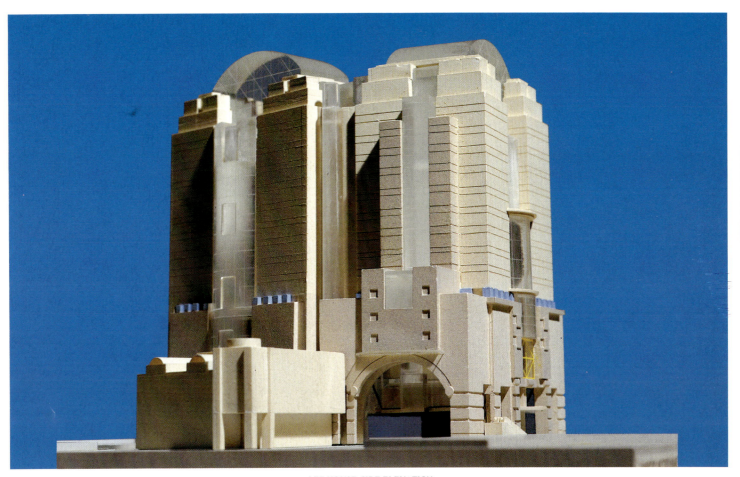

LEE HOUSE, SIDE ELEVATION

LEE HOUSE, MODEL OF INTERNAL STRUCTURE

LEE HOUSE, MAIN ENTRANCE

LEE HOUSE, MODEL BRIDGING LONDON WALL

NORMAN FOSTER

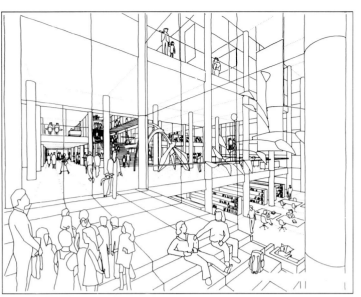

INTERIOR PERSPECTIVE ACROSS MAIN ENTRANCE

PERSPECTIVE VIEW OF MAIN STAIRCASE

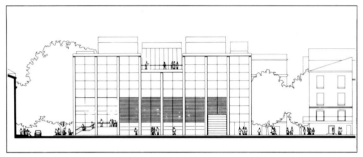

WEST ELEVATION

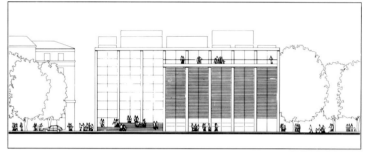

EAST ELEVATION

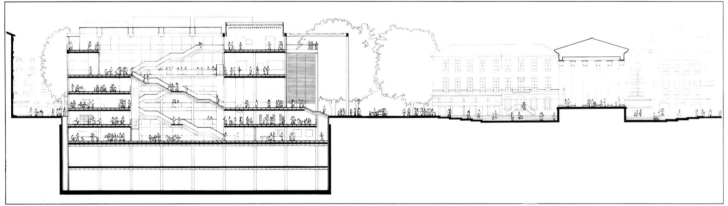

LONG SECTION THROUGH BUILDING SHOWING MAISON CARRÉE

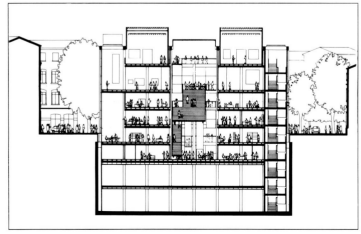

CROSS SECTION THROUGH BUILDING

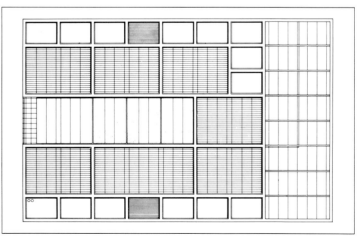

ROOF PLAN

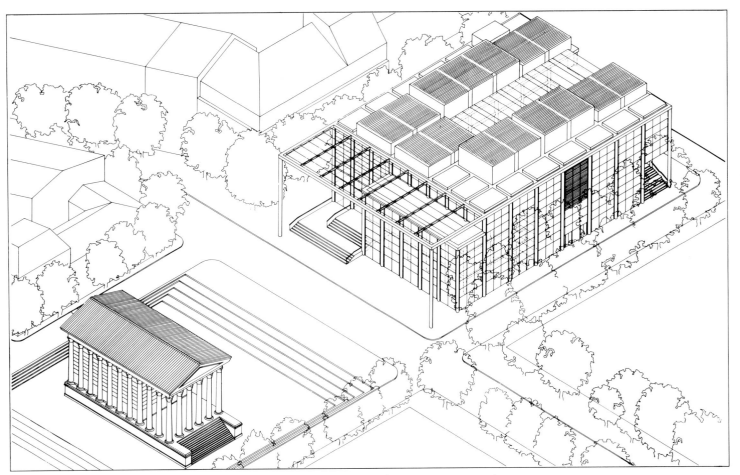

AXONOMETRIC VIEW OF MEDIATHÈQUE AND MAISON CARRÉE

FOSTER ASSOCIATES
Médiathèque, Nîmes

The site for the Médiathèque at Nîmes consisted of a car park and an Ionic colonnade which was the remaining facade of the old theatre constructed in 1803 and destroyed by fire in 1952. The site covers one half of the Place de la Comédie which is divided by the boulevard Victor Hugo. The other half of the square directly facing the Médiathèque site holds the Maison Carrée, the Temple of Caius and Lucius Caesar built in the reign of Augustus.

The new building will be culturally in the heart of Nîmes and the site taken as a totality will have a reference beyond the immediate boundary of its confines. The building recognises the importance of public space access and use, both in its internal public spaces and the exterior portico facing the Maison Carrée. It is a low building echoing the height of surrounding buildings and is referring in proportional relations to the temple facing it. In fact in relation to the context the building consciously avoids an industrial look, eschewing diagonals in the structure. The structure of the building comprises two levels of technology: glass and metal where the steel is either stainless or clad in bronze, but not painted,

and concrete and stone facing. A massive traditional shell integrates structural and environmental engineering.

The building is based on a 9m x 9m concrete grid structure and is extensively glazed externally with glass. It consists of three levels above ground and three below including a car park. The entrance attempts to unite both the literary and visual cultures represented by the building's library and gallery. The gallery for the permanent collection is at the top of the building in order to benefit from immediate access to natural light provided by the glazed roofs. Below the permanent collection are the galleries for temporary exhibitions which are lit by means of slots or light funnels that channel direct sunlight from above. Lighting is also provided by side light as well. The role of light within the building as a whole relates to the context of the site and town and assumes an almost poetic role during both day and night.

The Médiathèque representing the literary area of the building is positioned close to the ground. It consists of a reference library as well as cultural information facilities. The

three lower levels of the building comprise areas for stores, services and parking which formerly dominated the ground level of the site as well as a section of the square opposite.

The building is characterised by a route or atrium that runs through its centre. This fulfils a variety of functions, providing light, movement and orientation. A series of central ramps both suspended and stepped link basement and ground levels.

The facade of the building is distinguished by a large suspended canopy and stepped entrance which in effect defines a large symbolic and functional public portico. The site of the colonnade facade from the original theatre is now a large public area that has its own distinct character and that opens out into the square to provide a proscenium display area for theatre, film and music as well as trade displays that can be viewed from the other side of the square. Thus contextually the building provides a harmonious cultural and practical link uniting both sides of the square. Although its forms are distinctly contemporary the use of light and the purity of the structure blend sympathetically with the surrounding context.

MODEL VIEW FROM MAISON CARRÉE

MODEL VIEW OF NORTH ELEVATION

VIEW OF SITE TOWARDS MAISON CARRÉE

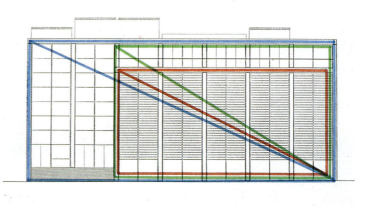

PROPORTIONAL STUDY OF MEDIATHÈQUE

AERIAL MODEL VIEW LOOKING TOWARDS MAISON CARRÉE

MODEL VIEW OF SOUTH ELEVATION

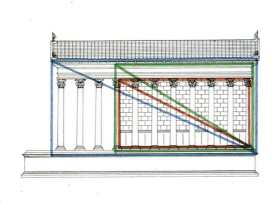

PROPORTIONAL STUDY OF MAISON CARREE

VIEW OF SITE FROM ABOVE MAISON CARRÉE

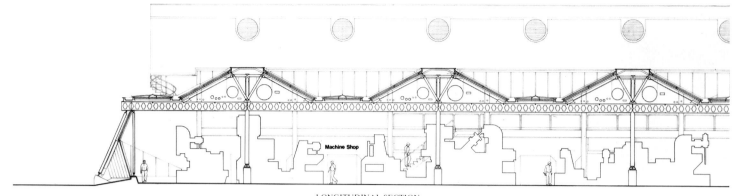

LONGITUDINAL SECTION

DETAIL OF RAKING PROPS MEETING

INTERIOR VIEW OF COVERED BRIDGE

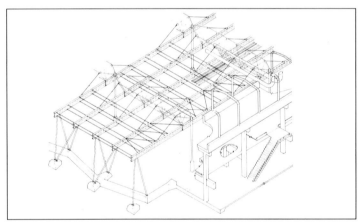

ISOMETRIC

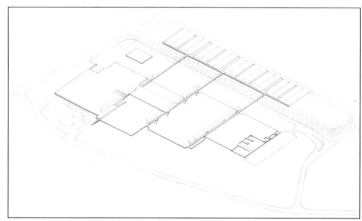

SCHEMATIC ISOMETRIC

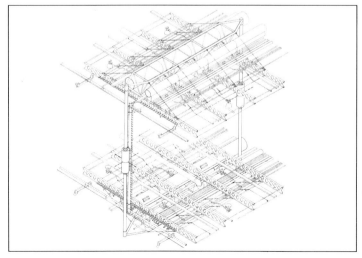

SERVICE CO-ORDINATION DRAWING

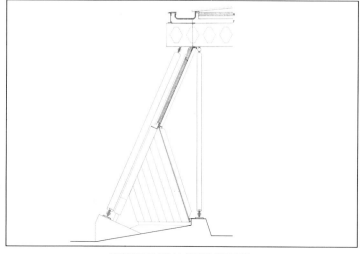

EXTERNAL WALL CROSS SECTION

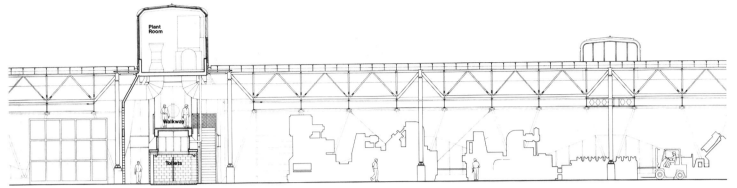

CROSS SECTION

GLAZED WALL OF PRODUCTION AREAS

MS TUBULAR RAKING PROP

AHRENDS BURTON & KORALEK
Cummins Engine Plant, Shotts, Lanarkshire

The redevelopment has taken place around and amongst the existing factory buildings in which Cummins have operated for the past 25 years and will have the capacity to produce ninety 250-400 HP diesel engines a day. The production areas have been divided into four distinct elements: Receiving, Machining, Stores and Assembly, and, lastly, Testing and Shipping. These four elements are placed in a progressive sequence, each serving the next. The planning allows for expansion of up to 30 percent in the future.

Above the east/west production flow there is a separate north/south pedestrian circulation system connecting to the car park at a higher level of the site. The car park stretches the full length of the factory allowing people to park near to their place of work. Three covered bridges provide direct access into the factory at the upper level, clear of materials movement. Stairs give access to locker, toilet and refreshment facilities below. The two principal bridges will be linked by an upper-level amenity deck which contains the Cafeteria, Medical Centre and Lecture Room. Part of the existing factory space is being converted into an open-plan office with a high degree of natural light.

New production areas of the buildings are structured on a 15 m² grid. Pin-jointed tubular steel columns support welded tubular steel primary trusses. Reinforced concrete stub columns at low level resist possible accidental impact loads from fork-lift trucks. Secondary roof structure takes the form of castellated steel joists fixed below the bottom member of the primary trusses. Steel rod hangers connect the primary trusses and castellated beams at one-third intervals across the 15m span. The roof profile follows this geometric configuration and provides a continuous zone for distribution of primary services. Secondary services distribute within the depth of the secondary structure thus establishing a clear set of differential zones which will maintain a coherent order for distribution routes in the future.

Cladding for roof and walls consists of an inner skin of corrugated structural steel decking (perforated for acoustic absorption), steel spacer purlins and mill-finished corrugated aluminium sheeting externally. Cavities are filled with thermal insulation material, vapour barriers and breather paper. Roof and wall glazing utilises green tinted glass and aluminium patent glazing bars.

Energy studies have been undertaken. The buildings are insulated thermally to standards well above the regulation requirements and utilise sophisticated air cleaning and ventilation systems to transfer heat from the high-heat production Machining Areas to adjacent parts of the building.

The new factory is built on a sloping site which resulted in a considerable amount of excavation into the hillside to provide a flat area. The surplus earth is formed into a mounded sculpture which serves not only as a windbreaker and a visual screen to the Receiving Area but creates a potent landmark on a scale appropriate to the new Cummins factory.

Collaborators: *Building Owner* Scottish Development Agency; *Client* Cummins Engine Company Limited; *Structural Engineers* Ove Arup & Partners; *Services Engineers* Ove Arup & Partners; *Quantity Surveyors* Monk and Dunstone Mahon and Scears; *Landscape Architects* Landesign Group.

ELEVATION SHOWING EXTERIOR CLADDING AND GLAZING

SPIRAL STAIRWAY LEADING TO COVERED BRIDGE

SPIRAL STAIRWAY

EXTERIOR ELEVATION

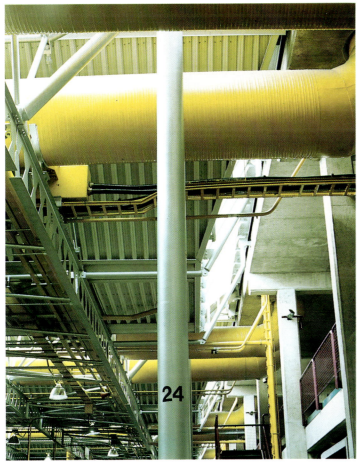

PIN-JOINTED STEEL COLUMN

COLUMN AND CASTELLATED SECONDARY BEAMS

PANORAMIC VIEW OF UNIFYING WALL

EXTERIOR DETAIL OF UNIFYING WALL

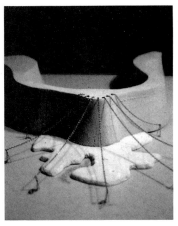  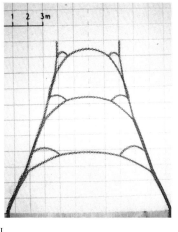

L TO R: MODEL OF CABLE TENT SUPPORTS: MODEL OF COMPLETE SCHEME: HANGING MODEL

CABLE TENT INTERIOR   HEART TENT INTERIOR

FREI OTTO
Diplomatic Club, Riyadh

The Diplomatic Club is intended to combine sporting, dining, recreational, conference and hotel facilities for the new diplomatic quarter of Riyadh. The club is characterised by lightweight cable net and fabric structures attached to a heavy limestone and masonry and concrete construction.

Originally conceived as a cluster of buildings united by means of a wall that enclosed a heavily landscaped interior garden, the scheme allowed the wall to develop the dual function of link element and building. This facilitated the control of temperature by means of heavy mass and the concentration of essential functions. A series of lightweight long-span structures attached to the heavy wall would then contain the large clear functional areas that would be used intermittently. It was proposed to construct the wall using local stone and the lightweight long-span structures as tents. The two tents facing the interior garden, for reception and banqueting purposes, would be of cable net construction with insulation and tile cladding, and would blend well with the intensive landscaping of the interior garden, as well as provide volume for formal functions which

could be environmentally controlled with relative ease. The tents facing outwards towards the plateau, for sports, restaurant and lounge purposes, would be of translucent fabric construction, providing a sharp contrast with the massive wall in the desert landscape.

The interior garden required a central feature, and a tent covered with glass tiles was proposed. Both PVC-coated polyester and Ptfe-coated cloth were considered for the fabric tents, the latter being chosen for its longer lifespan. Whilst PVC-coated polyester cloth is easier to fabricate, damages less easily during handling and can accept greater tolerances in erection and patterning, the PVC degrades in ultraviolet light. The constituent materials of PTFE-coated glass fibre cloth, however, are very stable and do not degrade with time, and so potentially have a long life, possibly 50 years or more compared with 10 to 20 years for PVC-coated polyester cloth.

The fabric tents are fundamentally conical in form and use radial supporting cables tied via A-Frame perimeter steel masts to the ground and to the top of the masonry wall construction via a fan mast. The fabric is

attached at the bottom boundary to a cable spanning between the perimeter masts and at the top boundary to a cable spanning around the fan mast. The fabric is attached to the wall using clamping plates to a roped edge tied back with a rigging screw type connection to a short length of rail bolted to the wall.

The heart tent is of cable net construction, using 6mm stainless steel strand at 326 mm centres. The boundary masts are tubular stainless steel tripods with a compression member and two ties. The 2,020 painted and toughened glass tiles are attached to the cable net by stainless steel clips.

Collaborators: *Owner* Gov of Saudi Arabia; *Architect* Frei Otto/Omrania; *Engineer* Büro Happold; *Contractor* Hanyang Corporation with specialist lightweight structures subcontractor, Stromeyer Ingenieurbau.

VIEW OF FAN TENTS ATTACHED TO EXTERIOR OF UNIFYING WALL

INTERIOR CABLE TENTS UNDER CONSTRUCTION

SECTION OF WALL UNDER CONSTRUCTION

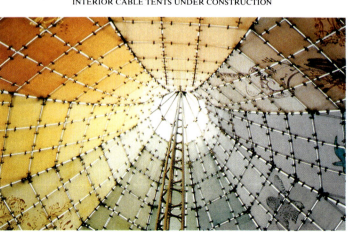

INTERIOR VIEW OF HEART TENT ROOF

FAN MAST OF EXTERIOR FABRIC TENT

VIEW TO HEART TENT IN INTERIOR GARDEN

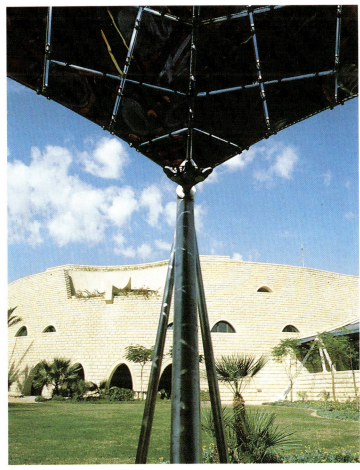

HEART TENT BOUNDARY MAST

INTERIOR VIEW OF UNIFYING WALL

DEREK WALKER

PIANO & ROGERS, ARCHITECTS/OVE ARUP & PARTNERS, ENGINEERS, POMPIDOU CENTRE

THE GREAT ENGINEERS: A SURVEY
Derek Walker

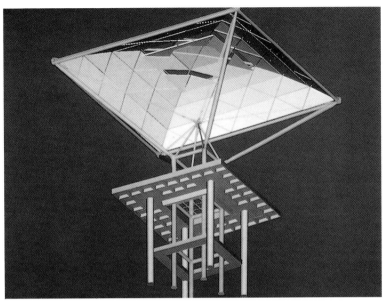

FOSTER ASSOCIATES. STANSTED AIRPORT. COMPUTER-GENERATED DRAWING
OF STRUCTURE AND SERVICE INTEGRATION

The following pages present, in encapsulated form, a survey of the exhibition on *The Great Engineers* which marks the 150th anniversary of the founding of the Royal College of Art. Then, in 1837, the industrial revolution was just getting into its stride. George Stephenson's 'Locomotive', the first locomotive to work on a public railway, had been built 12 years previously, and Isambard Kindgom Brunel was already at work on the Great Western Railway. The astonishing

variety and scale of the projects of these two men alone excite the imagination even today: the Britannia Bridge, the Royal Albert Bridge at Saltash, the High Level Bridge at Newcastle, the London to Birmingham Railway, the SS *Great Britain* and the SS *Great Eastern* to name but a few.

With the other engineers of their generation, Brunel and Stephenson re-shaped the landscape and environment of 19th-century Britain. His Royal Highness the Duke of Edinburgh pays homage to them in his foreword to *The Great Engineers*:

The [Industrial] Revolution was the creation of a number of remarkable engineers; many self-taught, whose energy and original ideas transformed life in this country and whose influence was to be felt throughout the world.

Just because engineering is a matter of manipulating materials does not mean that the great engineers of the eighteenth and nineteenth centuries were all unrepentant materialists. They were men of exceptional vision and imagination. Many of them made fortunes and most of them used their fortunes for the welfare of the community. Their names may be linked with their engineering master-pieces, but as individuals they were generous to charity, active in philanthropy, concerned with civic affairs and enthusiastic supporters of the arts.

Engineering was chosen as the theme for the Royal College of Art exhibition because it is a fitting reminder of the original intentions of the college's founders: namely to establish a school of design which would maintain links with industry – an aim which is as vital today as it has ever been. By including contem-

porary projects, it is also hoped to show that the spirit of great engineering and invention is still very much alive.

The text for this survey is extracted in the main from Professor Derek Walker's essay in *The Great Engineers*, 'Today's Engineers: The Legacy Lives On'.

High-quality structural engineering is a small world and the atelier process which develops the invention and elegance of technical solutions is, alas, confined to very few practices. The oldest established group is Freeman Fox, founded in 1857 by Sir Charles Fox, fresh from the dissolution of Fox Henderson of Smethwick. Fox and his partner were the real heroes of the Crystal Palace in carrying out for Paxton detailed design, fabrication and erection of the Exhibition Building in the unbelievably short space of eight months. Paxton was not Fox's only link with the heroics of the first half of the century; he also fulfilled every schoolboy's ambition by acting as an engine driver on the Liverpool-Manchester Railway for a short spell just prior to becoming a pupil of Robert Stephenson and, later, assistant engineer on the London-Birmingham Railway. It is perhaps appropriate that the prime linkage between the Crystal Palace and today's generation of engineers is bridged by this firm, which has been responsible for a wide range of major engineering works in over 30 countries. During the early years its work was largely concerned with railway projects including bridges, in Britain and abroad. Over the last 50 years activity has extended to other fields, including bridges, roads, expressways, tunnels, dams, hydro-electric and thermal power stations, buildings, special structures, radio and optical telescopes, mass freight and

The lithographs on this and the facing page were lent by Clive Wainwright

HAND-COLOURED DELAMOTTE LITHOGRAPH SHOWING CRYSTAL PALACE AFTER RE-LOCATION IN SYDENHAM

DELAMOTTE LITHOGRAPH OF CRYSTAL PALACE. BOTH THIS ILLUSTRATION AND THE ONE OPPOSITE SHOW THE ORIGINAL OWEN JONES COLOUR SCHEME

CLOCKWISE FROM L: GRAF ZEPPELIN *c*1928; BRUNEL'S SS *GREAT EASTERN*, 1858; FLOOR TO SLIP NO 4, WOOLWICH, 1870; BOILER ROOM, CHATHAM DOCKYARD

passenger transport systems.

From 1969, when the group carried out further studies and set up the network for the Hong Kong Metro, Freeman Fox have developed an in-depth know-how of mass and light transit railways, participating in projects in Iraq, Kuwait, Taiwan, Greece, the United Kingdom, Mexico and Venezuela as well as continuing an involvement in Hong Kong. This combined experience has placed the firm in the forefront of dealing with passenger transport systems.

The practice has also had a tradition of engineering works underground with the design and constructions of tunnels, particularly immersed tube tunnels. Douglas Fox was knighted for the Mersey Tunnel and many years later the practice completed the 6,000-foot cross harbour Hong Kong Tunnel. Of the 2,000 bridges designed by Freeman Fox, there are many memorable structures and a plethora of innovation and controversy.

Baptism by fire has always been a feature of the practice. Thus, when Ralph Freeman joined the firm in 1902 almost his first job was to make the calculations for the Victoria Falls Bridge. The project, completed in 1904 with a span of 500 feet, was the largest bridge of its type in the world. The Fox connec-

tion was maintained with Charles Beresford Fox, son of Francis, working on the team.

The tradition continued with the Sydney Harbour Bridge, the new Thames Bridge, the Forth Road Bridge, and the Severn and Wye bridges on the English/Welsh border. In recent years the Bosphorus Bridges and the Humber Bridge consolidated the large span development work pioneered by the practice. The Humber project in particular – the world's largest single-span bridge with a main span of 1,410 metres – is a masterpiece of design, economy and elegance.

In 130 years, Freeman Fox have maintained a reputation in all six continents, with railways from Cape Town to Rhodesia, the Snowdon Mountain Railway, the Argentine Central Railway, Victoria Falls Bridge, a definitive Report on the Channel Tunnel (in 1906), St Paul's Cathedral, Otto Beit Bridge over the Zambesi River, Auckland Harbour Road Bridge, High Marnham Power Station, Algonquin radio telescope, Ontario, Erskine Bridge, River Indus Suspension Bridge, Bangkok Expressway, Simon Bolivar Bridge over the Panama Canal and the Baghdad Metro.

Unlike Fox, Baker, Bazalgette and Williams, the giants of the

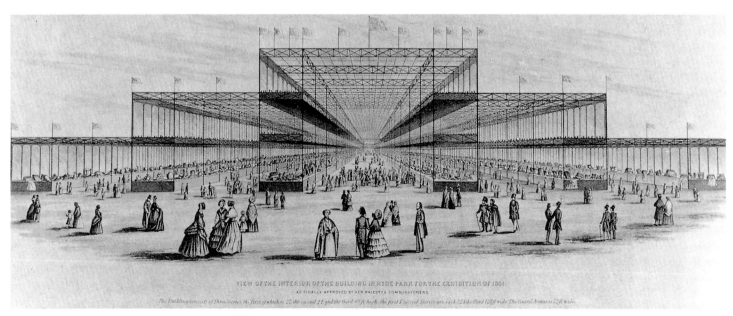

CRYSTAL PALACE, *c*1851. VIEW FROM *LONDON ILLUSTRATED NEWS*

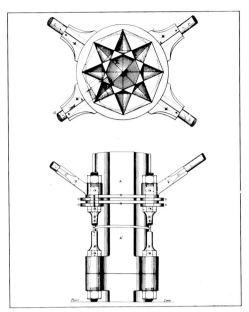
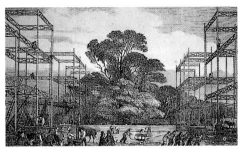

CRYSTAL PALACE, *L TO R*: DETAILS OF VERTICAL DIAGONAL BRACING BETWEEN COLUMNS; CONSTRUCTION VIEWS; HOISTING THE NAVE ARCHES INTO PLACE

post-war years were of Continental extraction. Ove Arup was from Denmark and Felix Samuely from Austria. Their paths briefly crossed in 1933 during Samuely's first post in England with J L Kier – the contractors in an engineering office under Ove Arup's direction.

Felix Samuely was born in Vienna in 1902. He studied in Berlin, where he obtained his engineering degree in 1923 and, apart from a year in an architect's office in Vienna, worked with contractors up to 1929 when he started his own practice with Stephen Berger. This lasted until 1931 when he went to Russia. During those two years Samuely designed the first commercial welded-steel building in Berlin, worked with Erich Mendelsohn and built a factory with Arthur Korn. On welding he writes:

> I have worked on the problem of welding of steel construction since 1928 when I had occasion to design an experimental, welded structure for a water cooler for Siemens Schukert. This tower was constructed as an imitation of a riveted building and it struck me forcibly that this was the wrong method of procedure and that welded steelwork should have its own design!

After two years in Russia working on the design of a steel-

works and researching steel construction, Samuely returned to Berlin via China and came to England in 1933. He obtained work with J L Kier contractors, his first job being the calculations for the Penguin Pool at London Zoo. He set up practice with Cyril Helsby after being asked by Mendelsohn and Chermayeff to act as engineer on the Bexhill Pavilion.

The practice prospered and Samuely was the engineer for many of the Modern Movement buildings of the late 1930s. He collaborated with such architects as Wells, Coates, Lasdun and Connell, Emberton, Goodhart-Rendel, Pilichowski, and Ward and Lucas on both steel and concrete structures. He was also an active member of the MARS group and in 1942, with Arthur Korn, became joint author of the *MARS Plan of London*. He had a passionate interest in transport and independently proposed his own rail plan for London.

Since Samuely's death in 1959 at the age of 57, the practice has been carried on by his brilliant protégé, Frank Newby, a Yorkshireman whose considerable skills have further extended the scope of the practice. Newby has also continued the tradition of collaboration on innovative building types, such as Leicester University Engineering Building with Stirling and Gowan; the

CRYSTAL PALACE, *L TO R*: 'OPENING CEREMONY' (LITHOGRAPH AFTER J NASH); VIEW DOWN THE NAVE (LITHOGRAPH AFTER OWEN JONES)

ROBERT STEPHENSON, BRITANNIA BRIDGE UNDER CONSTRUCTION, *c*1847

I K BRUNEL, RAISING THE TUBES AT SALTASH, *c*1864

JOHN HAWKSHAW, OLD CHARING CROSS STATION, *c*1864

W H BARLOW, ST PANCRAS ENGINE SHED WITH OWEN JONES' HOTEL COMPETITION DESIGN ON THE RIGHT, 1865

DEREK WALKER

L TO R: I K BRUNEL, GREAT WESTERN RAILWAY, FIRST TERMINUS AT BRISTOL, *c*1841; HAMPSTEAD ROAD CUTTING ON THE LONDON-BIRMINGHAM RAILWAY

Aviary at London Zoo with Lord Snowdon and Price; Cambridge Library, St. Andrew's Halls of Residence, Olivetti and the Tate Extension all with James Stirling; W D and H O Wills with SOM; in addition to projects as diverse as Bristol Cathedral, the US Embassy in Grosvenor Square and the Shopping Building at Milton Keynes.

Many other gifted engineers have worked in Samuely's office: they include Sven Rindle, whose Tatlin Tower will be remembered by all those who revelled in the Constructionist show at the Hayward in the early 1970s, and Anthony Hunt, who subsequently formed his own practice and engineered most of Team 4 and Norman Foster's earliest works with such panache – Reliance Controls at Swindon, Willis Faber at Ipswich and the Sainsbury Centre at the University of East Anglia.

Ove Arup, like Samuely, gathered around him exceptional designers; he has always praised and upheld the notion of multi-disciplinary work. He has been the Robert Stephenson of the 20th century: a designer who could organise and inspire simultaneously, capable of generating loyalty, dictating quality and delegating responsibility with the certainty and infallibility of a papal edict. Jenkins, Hobbs, Zunz, Beckmann, Michael, Rice,

Smythe, Lewis, Henkel and Barker are all superb designers who have continued to develop the depth and unique quality of the Arup organisation. Their great strength has always been an incessant zeal for research and invention, an attitude attracting brilliant graduates like bees to honey.

Ove Arup and his Partners have managed the impossible – the combination of high-quality work in a massive dispersed organisation. They have engineered many of the seminal buildings of the last 50 years – the Sydney Opera House, Pompidou Centre, HongKong Bank, Stuttgart Staatsgalerie – and they still find time for little gems like the Menil Museum, IBM's Travelling Museum, the Patscenter and the Durham Footbridge. They have provided a finishing academy for bright young graduates and have generously made their research facilities available to most architects interested in the extended learning curve of new construction techniques and environmental and material analysis. Their philosophy was adequately summed up by Ove and himself in a paper given at the Building Services Inaugural Speech at the Institution of Civil Engineers in 1972:

Architects and engineers both see themselves as designers.

And although the majority of engineers and a great number

L TO R: ROBERT STEPHENSON, BRITANNIA BRIDGE UNDER CONSTRUCTION; I K BRUNEL, GREAT WESTERN RAILWAY

74

L TO R: BRITANNIA BRIDGE UNDER CONSTRUCTION; I K BRUNEL, THE BOX TUNNEL, *c*1841

of architects can hardly be called that, it's the designers I am concerned with here. For the design, as I use the word, is the key to what is built; it is the record of all the decisions which have a bearing on the shape and all other aspects of the object constructed. These decisions are unfortunately not all taken by the designer but they must be known to him and integrated into a total design.

We must distinguish between routine design, which does not require any creative thinking, and what may be labelled original, innovative, conceptual or creative design. Creative design must of course build on previous experience and contains and employs pre-designed parts, and it may even consist almost entirely in assembling such parts to create an entity. But building is always tied to locality and to the people one builds for, and they vary from case to case. The synthesis required to create an entity, a whole which economises in means yet fulfils the aims, is an artistic process.

Art is solving problems which cannot be formulated before they have been solved. The search goes on, until a solution is found which is deemed to be satisfactory. There are always many possible solutions, the search is for the best – but there is no best – just more or less good. Quality is produced if the search doesn't stop at a second-rate solution but continues until no better solution can be found.

An engineer who doesn't care a damn what his design looks like as long as it works and is cheap, who doesn't care for elegance, neatness, order and simplicity for its own sake, is not a good engineer. This needs to be stressed. The distinctive features of engineering are mainly matters of content – the nature of the parts and the aims. The success of the whole undertaking depends on the right allocation of priorities and whether the resulting entity has this quality of wholeness and obvious rightness which is the mark of a work of art.

Arup's were also the engineers on the Sydney Opera House. A building that took 15 years to emerge from its chrysalis is worth more than passing reference, especially when its physical beauty still dominates the skyline of one of the most beautiful harbours in the world and where the poetry of its conceptual form has figured on every poster extolling the virtues of Australia since the Opera House opened in 1973.

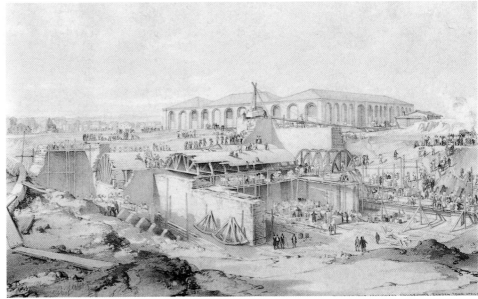

I K BRUNEL, GOVER VIADUCT, CORNWALL RAILWAY, *c*1858; ROBERT STEPHENSON, CAMDEN TOWN STATIONARY ENGINE HOUSE, *c*1836

UTZON; HALL TODD & LITTLEMORE, ARCHITECTS/OVE ARUP & PARTNERS, ENGINEERS, SYDNEY OPERA HOUSE

FOSTER ASSOCIATES, ARCHITECTS/OVE ARUP & PARTNERS, ENGINEERS, HONGKONG BANK

L TO R: SYDNEY OPERA HOUSE, CLADDING THE SHELLS; HONGKONG BANK, VIEW BY NIGHT

I remember attending a talk by Ove Arup when he was placed in an unacceptable dilemma, in the dark days just prior to Utzon's resignation. Whilst maintaining an unswerving love for Utzon's concept, which had seduced politicians and public alike, he was plagued by a growing awareness that Utzon was not prepared to face the problems of combining buildability and user requirements in the extraordinary silhouette which had become the Opera House.

The politics and confusion surrounding the sad saga left Arup's with many problems: the departure of a wayward genius, an escalating budget and a research and development requirement quite unprecedented for a building at that time. It was not the perfect scenario to complete the programme, especially when this was coupled with politics both fiscal and artistic which started to dominate progress.

The memory I retain of Ove in crisis is the clipped, no nonsense, slightly weary delivery as he turned around and around in his hand the wooden sphere illustrating the geometric order of the designs for the shells. Ove knew instinctively that the intuitive grasp of architectural and engineering fusion sat easily in his hands – the quality was going to be fine because the

old man said so! It was quite anachronistic, a re-enactment of the old virtues, the same logical, precise, professional, dispassionate concern for a good product.

Knowledge is still the engineer's solace in crisis, and fortunately Ove was able to tweak the invention and research needed to achieve solutions. The legacy of Stephenson, Paxton and Brunel surfaced instantly and Arup's did what they normally do when confronted with complexity and a seemingly endless series of problems – they solved them.

Sydney Opera House remains, affectionately, the built symbol of the 1960s.

Arup's have always survived in depth despite the loss of outstanding engineers over the years to academia and personal practice.

For 18 years Edmund Happold served Ove Arup and Partners as a design engineer and as executive partner for the innovative group of Structures 3 within the practice. Mannheim, Arctic City, the Council of Ministers in Riyadh, Pompidou and the hotels at Mecca and Riyadh came under his design initiative and, during that period, he developed a relationship with Frei Otto that has been outstanding in its consistently inventive

L TO R: OWEN WILLIAMS, BOOTS FACTORY INTERIOR, *c*1930-2; FOSTER ASSOCIATES, SAINSBURY CENTRE, EXTERIOR CLADDING

L TO R: LORD SNOWDON, CEDRIC PRICE, FRANK NEWBY, SNOWDON AVIARY, LONDON ZOO, 1962; BROWN DALLAS ASSOCIATES, OFFICE BUILDING, JEDDAH, *c*1980

development of lightweight structures.

Within Structures 3 a group of friends – Buckthorpe, Liddell, Dickson and Macdonald – started to develop the philosophy which was to become the rationale behind Büro Happold.

Since 1976 Büro Happold has worked in a bewildering variety of international environments: long-span structures, air-supported structures for Arctic conditions, airship frame design, bicycle design, buildings, tunnels and pneumatic flood control systems. If Arup is the Stephenson of the current period, Happold is the John Smeaton – versatile, passionate, a science-based artist with strong entrepreneurial talents. While President of the Institution of Structural Engineers, he used that platform to reactivate and renew the values of invention and leadership within the engineering profession. The sad reflection I have as an architect is that there are still too few engineers who can span the chasm that separates inventive design from humdrum mediocrity. This is not a factor in assessing the Happold group for they have assumed a personality well-adapted to the complexities of many of today's building types; looking at constructional problems in a multi-disciplinary way means that different members of the team assume leadership of a project at different times during its conception and implementation. For, in the final analysis, we should be producing buildings which reflect the times and they will never do this without an orchestra which can play in tune.

Perhaps it is appropriate to describe the unique quality of today's engineering by summarising Happold's personality – entrepreneurial, inventive, analytical, literate and lively. This is precisely how it was in 1837 and long may the legacy be maintained.

Nearly everyone in Britain, and perhaps elsewhere too, thinks that creative design in building is exclusively due to architects. But in building, the product is usually a complex one, requiring many skills inorder to put many values into it. I am a building or structural engineer working in partnership with architects and others, each group bringing a body of knowledge, experience and sensibility to a common problem. Today construction is about big money and to handle that successfully calls for toughness and rigour. Autocracy or selfishness are not called for, but a system of collective decision making is essential. For such a partnership means mutual authority and shared recognition amongst the members of the building team.

L TO R: FREEMAN FOX, THE HUMBER BRIDGE; OVE ARUP & PARTNERS, KINGSGATE FOOTBRIDGE, UNIVERSITY OF DURHAM, 1961-4

FREI OTTO, ARCHITECT/HAPPOLD & LIDELL, ENGINEERS, GARDEN CENTRE, MANNHEIM, *c*1975, CONSTRUCTION OF TIMBER LATTICE

FREEMAN FOX, THE HUMBER BRIDGE, ERECTING A BOX IN THE MAIN SPAN